HOPALONG CAS:

King of the Cowboy Mercha

HARRY L. RINKER

77 Lower Valley Road, Atglen, PA 19310

DEDICATION

Grace Bradley Boyd
and
Those for whom Hoppy was their Childhood Hero

Library of Congress Cataloging-in-Publication Data

Rinker, Harry L.
    Hopalong Cassidy: king of the cowboy merchandisers/Harry L. Rinker.
        p.   cm
    Includes bibliographical references and index.
    ISBN: 0-88740-765-X
    1. Cassidy, Hopalong (Fictitious character)--Collectibles--Catalogs. 2. Boyd, William, 1895-1972. I. Title.
NK808.R53   1995
791.43'651--dc20   94-49645
                                                              CIP

Published by Schiffer Publishing, Ltd.
77 Lower Valley Road
Atglen, PA 19310
Please write for a free catalog.
This book may be purchased from the publisher.
Please include $2.95 postage.
Try your bookstore first.

We are interested in hearing from authors
with book ideas on related subjects.

CONTENTS

INTRODUCTION

Hopalong Cassidy was my childhood hero. Like so many in my generation, I met Hoppy via television. Although an avid Saturday movie matinee goer in the late 1940s, my memories of that experience are entirely of the weekly serials, e.g., Commander Cody and Flash Gordon, not of the features and B movies that comprised the remainder of the bill. If I saw a Hoppy film or two, I do not remember. I wish I did.

Life without television-- ancient times. It seems so long ago, even to me. Yet, less than fifty years have passed since the arrival of the first television set on West Depot Street in Hellertown, Pennsylvania, where I grew up. Looking back, I cannot believe the tolerance shown by Gilbert and Mary Raudenbush as the neighborhood kids camped at their house on a daily basis to watch the B movie westerns and Captain Video. I do not remember exactly how much time passed before my parents or one of my aunts and uncles (two pairs lived in the same row) acquired a television set and relieved the pressure on the Raudenbushs. The period was probably months. In my memory, it was at least a year, possibly two.

Early television provided a wealth of cowboy heroes from which to choose. From the B cowboy movie genre came Rex Allen, Don "Red" Berry, Johnny Mack Brown, Sunset Carson, Larry "Buster" Crabbe, "Wild Bill" Elliott, Tim Holt, Buck Jones, Allan "Rocky" Lane, Lash La Rue, Ken Maynard, Tim McCoy, Tom Mix, Tex Ritter, Charles Starrett [the Durango Kid], Bob Steele, and the Three Mesquiteers [Stony Brook, Tuscon Smith, and Lullaby Joslin played by an ensemble cast that included at times Harry Carey, Ray Corrigan, Rufe Davis, Jimmy Dood, Ed "Hoot" Gibson, Raymond Hatton, Robert Livingston, Duncan Renaldo, Bob Steele, Max Terhune, Tom Tyler, and John Wayne]. My favorite was Hoot Gibson. I do not remember what I liked about him, only that I liked him. The problem was that Hoot's appearances were limited.

Childhood heroes need to visible. In the case of television, interaction with them should occur at least once a week, preferably more often. This left the big three - Gene, Hoppy, and Roy.

[**Note**: I realize that by limiting the number to three, I have relegated The Lone Ranger to secondary status. In my opinion, he deserves it. To paraphrase Senator Bentson's famous television debate remark to Dan Quayle: "He is no Gene, Hoppy, or Roy. Never will be. Never can be."]

I have never understood what anyone sees in Gene Autry or Roy Rogers in comparison to Hopalong Cassidy. They can sing, and Hoppy cannot. Big deal! In the areas that count - riding, shooting, and the manly arts - Hoppy was their equal *and more*. Further, he never fell in love and kissed a girl, at least not on the silver screen and television. He left that to his romantic sidekicks. Hoppy kept his virtue and purity intact.

Do you realize what I have done? I have written about Hoppy as though he was a real person - alive and breathing just like Gene and Roy. Hoppy is not real. He is a fictional character, identical to Mickey Mouse or Bugs Bunny.

William Boyd, the actor who portrayed Hopalong Cassidy, is the real entity. What makes Hoppy and Boyd's portrayal of him unique is that Boyd was able to merge his personality into that of Hoppy, thus creating an entire generation of post-World War II children that saw Hoppy as real. Hopalong Cassidy is one of the few fictional characters to achieve this distinction.

I remember exactly when I realized the dichotomy between William Boyd the person and Hopalong Cassidy, the fictional character. It happened in the mid-1970s when I was reviewing the television cowboy hero images on a new calendar that I had ordered for the upcoming year. Across the top of each page was the name of a major cowboy actor and sketches relating to the characters and roles that they made famous. Eleven of the individuals were real people, e.g., Roy and Gene. Clayton Moore represented The Lone Ranger. However, the twelfth name was Hopalong Cassidy, not William Boyd. Boyd was mentioned in a footnote.

I grew up believing Hoppy was real. Even as a youngster, I knew better. I did not care. I am not certain I care now There are moments when Hoppy is as real today as he was during my early teens. Somehow, it just seems right.

Analysis takes the fun out of things. Is not the fact that I preferred Hoppy over Gene and Roy enough? Do I have to explain why? Yet, I feel an explanation is necessary. So much for the mysteries of life.

My main attraction to Hoppy was his image as a friend who possessed the insights, wisdom, patience, and tolerance gained only through an extensive life time of experience. Hoppy used brains, not brawn to solve the puzzles that he and his companions faced. Roy and Gene were more like "good-old-boy" big brothers. They still had some growing up to do. As the first born in my family, I had no need or desire for a big brother, no matter how neat he may have been. The role of friend and father figure usually feel to the comic sidekick in Gene's and Roy's films. In real life, it belongs to the leader.

Hoppy arrived in my life at a time when my interest in the opposite sex was nil, minuscule if you want to be charitable. Life on the open range where a man in the companionship of a group of true-blue, time-tested companions battled the elements had great appeal. Women only complicated and clouded the picture. Watching a Gene or Roy movie provided more than ample proof of this fact. I always thought the reason Lucky was so lucky was that he ended most of his romantic entanglements by the picture's end.

My high church Lutheranism and fanatical devotion to the Boy Scouts with its laws and promises made me an ideal candidate to accept the strong ethical and moral code of living incorporated in Hoppy films. There is no question that Hoppy was one of the strongest influences in my life. I stick to my guns when I think that I am right. I fight for the principles in which I believe no matter how overwhelming the odds. While appearing on the surface as a relatively uncomplicated character, Hoppy wheeled and dealed, always keeping a trick or two up his sleeve. Makes sense to me.

Hoppy had a vision. He recognized that the west as he knew it would give way to a new era, one in which education would play a major role. I listened, believed, and acted accordingly. As a result, I became a 1950s nerd. I am not certain Hoppy would have fully approved; but, he respected individual choice, and I rest easy with that.

Although William Boyd died in 1972, Hopalong Cassidy lives in the hearts and minds of many of the men and women of the fiftysomething plus generations, including my own. Tragically, the current generation of youngsters and young adults has not been exposed to Hoppy and the principals for which he stood. The challenge is to make Hoppy live again. The society of the 1990s could learn much from heroes of Hoppy's stature and principles.

Not many individuals have an opportunity to pay back part of the debt they owe to someone who shaped their lives so profoundly. I am one of the lucky ones. This book is partial payment, as were the Hoppy, Gene, and The Lone Ranger exhibition at the Gene Autry Museum of Western Heritage in 1992 and the Hopalong Cassidy exhibition at the American Museum of the Moving Image in 1994-95 in which I participated. If you look closely, you will find Hoppy and/or Hoppy collectibles mentioned and/or pictured in many of my books and articles. I am doing my best to keep Hoppy's memory alive.

Fortunately, I am not alone. A Hoppy revival is in the wind. An annual Hopalong Cassidy Festival is held each year in Cambridge, Ohio. U.S. Television Offices, owners of the character rights to Hopalong Cassidy, are licensing new products, working actively to return Hoppy to television, releasing on video full length versions of the Hoppy movies and television programs, finalizing plans for a new Hoppy movie, and involved in discussions for a Hopalong Cassidy theme park. As this book goes to press, Hoppy is king-of-the-hill in the area of television cowboy hero collectibles, despite his current lack of exposure.

Will this wealth of activity create a new generation of Hoppy devotees and collectors? Nothing would please me more. Yet, I cannot help but wonder if 1990s realism precludes the possibility of any revival of a romanticized west and the larger than life cowboy heroes associated with it. Gene, Hoppy, Roy, and their fellow range riders of the silver screen and early television blended reality, myth, and idealism as though it was a single life giving force.

I am glad that I grew up during a time when that convergence could be accepted without question. To hell with reality. Indulge in a little healthy, romanticized idealism. Take a few minutes out of your reality filled life and watch a few Hoppy movies on video. Immerse yourself in the genre - its imagery and philosophy. If it does nothing for you, at least it will help you understand your parents and grandparents better. That in itself makes the exercise worth while.

ACKNOWLEDGMENTS

A procrastinator by nature, I delayed the completion of this project longer than any other I have undertaken, frustrating several friends and more than one potential publisher. I know why I did this. However, this is one case where knowing the problem did little to solve it.

For years I have lived with the fear that completing this book will mark an end to the special bond that exists between Hoppy and myself. I convinced myself that as long as the final chapter remained unfinished, I was safe.

Time to bite the bullet and finish. I confess to being scared to hell as I prepare to ride the unchartered range ahead.

First, thanks to U.S. Television Office, Inc., owners of the literary, character, and other rights to Hopalong Cassidy. They provided the funds and much of the encouragement that resulted in this project reaching fruition. Jerry Rosenthal generously gave of his time to provide video tapes of the Hoppy movies and television shows, recordings of the radio shows, and key material from the company's archives. Holger Wrede not only became the vehicle that drove this project to conclusion, but, one of the major Hoppy collectors in the country. Judge Edward Fiala served as the voice of patience and calm.

Second, no one endured the trials and tribulations involved in weaning this manuscript from my mind more than the staff at Rinker Enterprises, Inc. - Ellen Schroy, Terese Oswald, and Dana Morykan. They helped organize and type much of the appendix material, quietly suffered verbal abuse that should have been more correctly directed at me, and freed me from other responsibilities so that I could *once again* get the damn manuscript done. I do not think they believed the day would ever come when the project finally would be finished. Harry L. Rinker, Jr., my son and Art Director at Rinker Enterprises, Inc., did the photography. My deepest appreciation to them all.

Third, In the course of preparing this manuscript, I participated in the planning and implementation of two major museum exhibits focusing on Hopalong Cassidy collections. Information uncovered by museum researchers and curators during the course of these exhibitions was shared with me. Thanks to Joanne D. Hale, James H. Nottage, Mary Ellen Nottage, John Langellier, and Cynthia Harnisch of the Gene Autry Western Heritage Museum of Los Angeles, California, and to Rochelle Slovin, Richard Koszarski, and Shira Nichaman of the American Museum of the Moving Image of Astoria, New York, New York.

Fourth, the Western hospitality and professionalism extended to me by the members of the American Heritage Center of the University of Wyoming was most welcomed. During my research there, they responded willingly and graciously to every request that I made, from an inordinately large number of photocopies to 35mm color transparencies of numerous items.

Fifth, in a never ending battle to extract the final pages of this manuscript from me, Peter Schiffer, publisher of Schiffer Publishing, promised to create a Harry L. Rinker Delinquent Author award in the book publishing field in my honor. I think he should. Heavens knows I set high standards. At the same time, Peter should establish a second award -- The Patient Publisher Award. I know it is tacky to create an award and then give it to yourself, but, he certainly is my nominee. The actual publishing chores for this book were the responsibility of Douglas Congdon-Martin. My thanks for the skills and persistence, he brought to the task. The fine design of the book is because of the skill of Ellen J. (Sue) Taylor.

I have saved my final thanks for two extraordinary groups - those who grew up with Hoppy and accepted him as their cowboy hero and those present day collectors and dealers of Hopalong Cassidy material. Each group shared personal remembrances, provided access to Hoppy material in their possession, and tolerated my many follow-up requests. This book is for them and about them. Any attempt to list them all inevitably will result in several individuals being missed. As a result, a great big HOPPY THANKS to everyone.

"Border Vigilantes" featuring William Boyd, with Russell Hayden, Andy Clyde, Victor Jory, Morris Ankrum, Frances Gifford, and Brit Wood.

Chapter 1

THE EVOLUTION OF HOPALONG CASSIDY

During the first two-thirds of the twentieth century, three different generations grew up with Hopalong Cassidy. The oldest, the generation immediately preceding and following World War I, read about Hoppy through the novels and short stories of his creator, Clarence E. Mulford. The next generation, those whose childhood occurred immediately preceding and following World War II, knew Hoppy from the movies. The final group, youngsters whose childhood spanned the late 1940s through the mid-1950s, identified with the television Hoppy. Novels, movies, and television - three very different mediums and three *very different* Hoppys.

Hoppy's chronological appearance in three different mediums - novels, movies, and television - suggests the possibility that the character evolved when and as the medium changed. Such an assumption is false. Each medium produced a Hoppy unique unto itself. Further, Hoppy grew and developed within the confines of each medium. The Hoppy that one meets at the beginning of the novels, movies, and television era is not the same Hoppy one meets at the end. There were three *very different* Hoppys, each serving the needs of their generation.

Hoppy's evolution in the novels was the most traditional since it was based on the chronological development of the human character. A brash young Hoppy was transformed into a Western sage. Hoppy's evolution in the movies occurred on two levels: (a) William Boyd's transition from a silent to a talking film actor and (b) the philosophical approach to the character in the movies produced by Harry "Pop" Sherman versus the approach in those produced by William Boyd. When the movie era ended in 1948, William Boyd still had an identity separate and apart from Hoppy. There were some signs of the merging of the fictional character and William Boyd into one, but nothing concrete. It is television that turned Hoppy into a live human being. The reel Hoppy became a real Hoppy. While most individuals think of the television Hoppy as a single entity, there were actually three Hoppys on the television screen at the same time - the movie Hoppy (cut down versions of the full length Hoppy films were shown), the television Hoppy (a character developed in the half-hour, made-for-television programs), and the idealized Hoppy (an imagined character that Boyd made real by living him in public as well as on television).

Hoppy of the Novels

Which of the following are not characteristics of Hopalong Cassidy -- drifter, drinker (occasionally to excess), two-fisted fighter, gunman (never hesitating to use a gun, even to the point of killing someone, when the occasion demanded it), prankster, tobacco chewer, vigilante (quite willing to take the law into his own hand), and womanizer (as a youth)? None of the above, some of the above, or all of the above? The answer is *all of the above*. If you answered incorrectly, its time you read a few or all of the Clarence E. Mulford novels in which Hopalong Cassidy appears.

Mulford's Hopalong Cassidy was a standard 1920s and 30s Western novel cowboy hero. He had faults. Who doesn't? He showed his character by what he learned from them. Hoppy was the embodiment of the agrarian myth - a man who pitted his skills against nature, winning on some occasions, losing on others, never quitting, always trying to move forward.

Hopalong Cassidy was in his early twenties in *Bar-20* (1907). By *Hopalong Cassidy Serves A Writ* (1941), he was a man who had lived life to the fullest, accepted the cards fate had dealt him, and waited patiently to participate in the last great round-up. The Mulford novels chronicle Hoppy's development from brash youth to elder statesman. The story is not told chronologically. There are occasional flashbacks.

The key to understanding the Mulford Hoppy is to examine the characteristics of the mature Hoppy, the Hoppy that appears in *The Coming of Cassidy* (1913) and later novels. Hoppy is the great western generalist. His skills are universal and honed. He is versed in every needed survival skill. In the group that surrounds him, there always is someone who is a better rider, shooter, tracker, etc. However, these are single-talent individuals. Hoppy's multi-talented character provides him the luxury of drawing upon these individuals' skills when needed, knowing full well that if they were not available, he probably would succeed just as well.

Mulford's Hoppy is a leader, but not in the traditional sense. He teaches by example without explaining what the example is or that the teaching process is occurring. Hoppy takes individuals under wing, but sends the fledglings out of the nest as soon as he feels that they are able to fly. His principal teaching methodology is learn and grow from your own experiences.

Mulford's Hoppy has a strong degree of self-reliance. He proves time and time again that he can survive on his own. His independence is never at question. It is this very self-confidence that allows him to work well within a group. He knows that others trust and rely on him. Likewise, he surrounds himself with individuals whom he can trust to hold up their end.

Little has been written about two aspects of the Mulford Hopalong Cassidy character that very much shaped and molded him: the married Hoppy and Hoppy as a loner. Yes, Hoppy was married in the Mulford novels. Not only was he married, he had a son. In *Buck Peters, Ranchman* (1912), Tex Ewalt is visiting the Bar-20 and asks Lanky: "'Where's Hopalong?,' demanded Tex. 'Married! H--l he is!' A strange look flitted across his face. 'Well, I'm d--d!...'"

The following exchange occurs a few pages later and provides a glimpse of Hoppy's wedded bliss and interaction with his peers:

"Mary, dish in hand, paused between the stove and the table. She looked at Tex with mischievous eyes: 'Billy-Red tells me you love him like a brother. Is he deceiving me?'

"Hopalong laughed and Tex replied, smiling: 'More like a sister, Mrs. Cassidy.--I can't find any faults in him, an' we don't fight.'

"Mary completed her journey to the stove, filled the dish and carried it to the table; resting her hands on the edge of the table, she learned forward in seeming earnestness. 'Well, you must know that we are one, and if you love Billy-Red ---' finishing with an expressive gesture. 'Those who love me call me Mary.'

"Tex's face was gravely wiseful, but a wrinkle showed at the outer corner of his eyes. 'Well,' he drawled, 'those who love me call me Tex.'

"'Good!' exclaimed Hopalong, grinning.

"'An' I'm thankful that my hair's not th' color to cause any trustin' soul to call me by a more affectionate name,' Tex finished. He ducked Hopalong's punch while Mary laughed a bird-like trill that brought to her husband's face an expression of idolizing happiness and made Tex smile in sympathy..."

A domesticated Hopalong Cassidy - who would believed it? The situation was too good to last. A married Hoppy restricted Mulford's character development. Two novels later, in *The Bar Twenty-Three* (1921), Mary and their young son were dead. Red Conner, his closest friend, explained: "He lost his wife an' boy a month ago - fever - in four days. He's all broke up. Went loco a little, an' even came near shootin' me because I wouldn't let him go off by himself. I've had one gosh-awful time with him, but finally managed to get him headed this way by talkin' about Johnny a-plenty. That got him, for th' kid allus was a sort of son to him."

From this point forward, Mulford's Hopalong Cassidy was a drifter. He moved in and out of his friends' lives as they moved in and out of his. He adopted numerous troublesome youths, setting them on the straight and narrow before turning them loose on their way. Yet, despite the bonds of loyalty and companionship that developed from these interactions, Hoppy was very much alone -- a loner. His companions could get close, but never too close. He was a person who had to have breathing room.

The Hopalong Cassidy character in the final Mulford novels has come to grips with and accepted the harshness and terms of life in the American West. It is the struggle that is important. The measure of a person's worth is how he reacts and grows from that struggle.

Some have suggested that Mulford's Hoppy was engaged in a Darwinian struggle - survival of the fittest, nature's noble man. While it is true that there is something ennobling about Mulford's Hoppy, there also is something very ordinary. In the final analysis, Mulford's Hoppy neither won nor lost. He survived. In the Western novels of the 1920s and 30s that in itself was a major accomplishment.

Learn More: Bernard A. Drew, *Hopalong Cassidy: The Clarence Mulford Story* (Metuchen, NJ and London: The Scarecrow Press, Inc., 1991); Francis M. Nevins, Jr., *Bar-20: The Life of Clarence E. Mulford, Creator of Hopalong Cassidy with Seven Original Stories Reprinted* (Jefferson, North Carolina, and London: McFarland & Company, Inc.: 1993).

Hoppy of the Movies

Hoppy appeared on the silver screen five years before the last Hoppy Mulford novel was published. After Harry Sherman obtained the film rights to the Hopalong Cassidy books in 1935, he did not invite Mulford to participate in the process of moving Hoppy from page to screen or to come to Hollywood to serve as an advisor on the set. Sherman planned to produce stock Westerns, loosely adapting the Mulford characters and story lines to the established Hollywood genre for the medium.

What did Mulford think of Sherman's movie Hoppy? In Rudolph Elie's "The Roving Eye" column in the October 26-27, 1950, *Boston Herald*, Mulford is quoted as saying: "Why, it's absolute nonsense. If Hoppy ever showed up in a saloon in duds like that they'd shoot him down on sight. Hoppy was a cripple, but I taught him everything he knew. I taught him how to take cover, I taught him to be brave without being foolhardy, I taught him how to handle liquor, I even taught him how to be decent with women...!" In *Hopalong Cassidy Takes Cards* (1937) and *Hopalong Cassidy Serves A Writ* (1941), the last two Mulford Hoppy novels, both of which were written after Hoppy was established on the silver screen, Hoppy is portrayed exactly as Mulford portrayed him in his earlier novels. Mulford made no attempt to change his character to the romanticized, sanitized Hoppy of Harry Sherman's films.

The Hopalong Cassidy films divide into two distinct groups - those produced by Harry Sherman and those produced by William Boyd Productions (the last twelve films, a.k.a., the North Vines). Before focusing on the evolution of the Hoppy character, two attributes of the films, most notable during the Sherman era, deserve attention These are the concept of the trio Western and spectacular production values.

If Harry Sherman did not invent the concept of the trio Western, he certainly perfected it. By the end of the 1930s, the trio Western - hero, romantic sidekick, and comic sidekick - became the industry standard. While Roy, Gene, and a few others combined the hero and romantic sidekick in one, they could not ignore the public's clamor for the comic sidekick. Although a review of the Hopalong Cassidy filmography reveals an expanded list of romantic leads and sidekicks, there are two romantic leads, Russell Harlan and Rand Brooks, and two sidekicks, George Hayes and Andy Clyde, that are most synonymous with the Hopalong Cassidy movies.

[**Note**: Here is a trivia question for Hoppy fans? Who played the romantic sidekick in the earliest Hoppy movies and who played the comic sidekick in the made-for-television programs? Stumped? Jimmy Ellison played Johnny Nelson in the first series of Hopalong Cassidy movies. Edgar Buchanan was the comic sidekick in the television series.]

Film buffs constantly rave about the production values of Sherman's Hoppy films - above average screen plays, lavish locations, rich stock footage, and quality editing. A typical B movie Western of the era was shot in one week. Sherman took two. Sherman provided his directors and production staff with a generous, albeit tightly controlled, budget that allowed for the extras - additional personnel (both in front and behind the camera), extensive on-site shooting, and time to edit the material properly.

Film buffs view the North Vine Hoppys, i.e., those produced by William Boyd Productions, in a lesser light. Budgetary constraints hindered Boyd. However, the North Vine films allowed Boyd to develop the Hopalong Cassidy character as he envisioned him, not as Sherman viewed Hoppy. If one looks carefully, one can see the television Hoppy beginning to emerge.

The Sherman Hopalong Cassidy films typecast Boyd, a fact he recognized by the early 1940s. In hindsight, Boyd's portrayal of a single character in a sixty-six film series is remarkable. In 1995, series films about Batman, Dirty Harry, Star Trek, Superman, and the Terminator still have not passed the tenth film. The only series that has surpassed this mark is the ghoulish Halloween thrillers. While Roy and Gene made more films, they played themselves or some fictionalized character of the moment. Their work does not constitute a series.

There is little question that the character of Hopalong Cassidy evolved significantly in the Hopalong Cassidy films. The most obvious evolution is in the acting and filming techniques. The silent era, an era in which William Boyd was a major star, had just ended. It makes perfect sense to assume that Boyd would bring the acting techniques and facial expressions that had served him well in silent films to the talking screen. The same holds true for Hoppy's producers, directors, and the camera crews. Evidences of silent era influences abound in the Hoppy films of 1936 and 1937. Everyone was learning a new craft. By the late 1930s, everyone was comfortable on a sound set. Boyd knew how to use the camera to his best advantage. The Hoppy of the 1940s films was an individual whom one accepted as naturally as a best friend or next door neighbor.

Much has been made of the black and white imagery of the Hoppy films - colors that suggest a constant struggle between good verses evil. Whether or not it was deliberate or a lucky choice, Sherman recognized its value immediately and capitalized on it. It was a prevalent, but hidden theme of the television Hoppys. In the black and white Cold War era, no one was interested in shades of gray.

Hopalong Cassidy limped only in the first film. After that, Hoppy walked straight and sat tall in the saddle. His novel signature attribute had little value in the movies.

Where as Hopalong Cassidy matured and aged during the course of the novels, the films focused on Hoppy as the mature, Western sage. His character had a benevolent Orwellianism - all cowboys are created equal, but some cowboys are more equal than other cowboys. Hoppy was not an equal, nor was he a brother, father, or grandfather figure. He was a living legend. If the West had gurus, he would have been one.

Once again, whether a deliberate decision was made to

"Secret of the Wastelands," featuring William Boyd, with Brad King, Andy Clyde, Barbara Britton, Douglas Fowley, Keith Richards, Soo Yong, Gordon Hart, and Hal Price.

sanitize Hoppy or it was a response to the public reception to the character as initially portrayed is not important. The truth is that by the late 1930s Hopalong Cassidy became an symbol of idealized Western moral virtue. He drank sarsaparilla not liquor, uttered no foul language, never used his guns or fists when a problem could be resolved peacefully, and maintained a keen sense of independence, especially when it came to members of the opposite sex. No wonder Mulford did not recognize him.

One of the key differences between the movie Hoppy and the television Hoppy is the use of Hoppy as a teacher of morality. Ample evidence exists to suggest that William Boyd's vision of Hoppy was vastly different from that of Sherman. Boyd saw in Hoppy a character with the potential to shape and mold youth. He wanted to portray him that way. Sherman did not.

When William Boyd Productions acquired the movie rights to make new Hoppy movies along with the theatrical re-release rights to all but the first six of the Hoppy movies that Sherman had produced, Boyd finally could shape Hoppy to his image. Boyd moved cautiously. The changes had to be subtle. The Sherman Hoppy was clearly established in movie goers' minds.

Boyd's Hoppy was teacher, father confessor, seer, and sage - all rolled into one. While he led by example, the lessons were much more obvious. Whereas Sherman's films were heavily adult oriented, Boyd's films were focused toward a much younger audience. The twelve North Vine films were only moderately successful. If Hoppy was to continue riding the range, he would have to find a new audience. Enter television.

Learn More: Francis M. Nevins, *The Films of Hopalong Cassidy* (Waynesville, North Carolina: The World of Yesterday: 1988).

The Television Hoppy

Identifying the Hopalong Cassidy character that evolved between 1948 and 1952 as the television Hoppy is a misnomer. The merchandised Hoppy or the real Hoppy are more correct technically. The television Hoppy is used because the evolution occurred after the time period when Hopalong Cassidy appeared on television.

After 1948 Hoppy became real. William Boyd made a conscious decision to suppress his own personality and become the personification of the fictional character he had portrayed since 1936. Whenever he appeared in public, Boyd and Hoppy were one. When people addressed Boyd as Hoppy, a common occurrence, he responded. Even his intimate friends called him Hoppy.

William Boyd and Grace Bradley Boyd never had children of their own. Like his fictional counterpart, Bill Boyd did not let this hinder him. He adopted an entire generation of children between the ages of five and fifteen. Boyd as Hoppy took upon himself the task of providing them with a set of values that served as the foundation for their adult life.

One of the key concepts in Boyd's approach was to treat children not as youngsters, but as young adults. Some may argue that the approach is Victorian. To some degree, it is. However, it also corresponds to the Horatio Alger optimism that dominated the late 1940s and early 1950s. Success awaited anyone who applied themselves and followed traditional paths.

The essence of Boyd's approach is summarized in an eight-point *Hopalong Cassidy Troopers Creed*. It states:

To be kind to birds and animals
To always be truthful and fair
To keep your self clean and neat

To always be courteous
To be careful when crossing the streets
To avoid bad habits
To study and learn your lessons
To obey your parents

Gene and Roy also had cowboy creeds. However, their creeds consisted of ten points, rather than eight. Ten points, ten commandments. It does not take a genius to see the connection. Boyd countered with a ten point *Hopalong Cassidy's Creed for American Boys and Girls*. Nowhere near as succinct as the *Troopers Creed,* it reads:

1. The highest badge of honor a person can wear is honesty. Be truthful at all times.
2. Your parents are the best friends you have. Listen to them and obey their instructions.
3. If you want to be respected, you must respect others. Show good manners in every way.
4. Only through hard work and study can you succeed. Don't be lazy.
5. Your good deeds always come to light. So don't boast or be a show-off.
6. If you waste time or money today, you will regret it tomorrow. Practice thrift in all ways.
7. Many animals are good and loyal companions. Be friendly and kind to them.
8. A strong, healthy body is a precious gift. Be neat and clean.
9. Our country's laws are made for your protection. Observe them carefully.
10. Children in many foreign lands are less fortunate than you. Be glad and proud you are an American.

Note the strong emphasis on family values. One of the keys to Hoppy's success in the 1950s was strong parental support. Hoppy's values were their values. Take a few minutes and read the Trailers in Appendix #3. Here are two abbreviated examples:

"Hi, little partners...Have you been doing anything to help Mom around the house lately? You haven't? Let's do all we can to help, eh? I'll bet you have fun doing it - and I know Mom will appreciate it. Would you do that for me?..."

"...Now here's a little message for my little friends...always mind your Daddy and Mommy. Please remember, they're the best friends you have in the world..."

Boyd made certain that Hoppy never challenged parents for their children's trust, loyalty, and confidence. Hoppy was a friend, not a father figure. His role was supportive, not dominate. By living an exemplary life that embodied traditional family values, Boyd's Hoppy provided a model to which parents could point.

Deciding to make Hoppy a figure for moral good was only the first step. If the concept was to work, Boyd had to take Hoppy to his fans. He did. In the early 1950s Boyd engaged in a series of public appearances at department stores, parades, fairs, and civic events that would have taxed a much younger man. At the time, Boyd was in his early fifties.

There is no question that Boyd was driven. He wanted to meet and greet as many of his fans as he could. Often this meant standing in place for eight hours and longer. As long as there was a hand to shake, he was going to shake it. When interviewing the adults and children who stood in these lines to shake Hoppy's hand, the one thing they always comment about is the sensation of warmth they felt when Hoppy's hand grasped their's. Almost everyone claims to have had a near mystical experience.

Much has been written about how Hoppy merchandising made Boyd rich, a millionaire several times over in the space of a few short years. True. However, not enough has been written about Boyd's recognition and understanding of the source of his wealth and his desire to give something in return.

One little known fact is that while Boyd accepted expense payments for his appearances, he never kept a public appearance fee. These fees always were donated to a local charity, preferably a children's hospital if there was one in the area. Hospital visits were a common method used by Hollywood stars to gain a little extra publicity. They were a commitment with William and Grace Bradley Boyd. Hoppy visited a hospital whenever the opportunity presented itself, his focus, the children's ward.

Further, Boyd felt strongly that youngsters should not have to pay to see him. Early in his department store appearances, several stores tried to institute a policy that a child had to purchase an article of Hopalong Cassidy merchandise in order to get in line to shake Hoppy's hand. The policy was abandoned the moment Boyd heard about it. No one would be allowed to put a dollar value on Hoppy.

While television provided the exposure that Boyd used to propel the character of Hopalong Cassidy into public dominance, it had virtually nothing to do with the evolution of Hoppy during this period. Initially Boyd released slightly cut down, one hour versions of the Hopalong Cassidy films to television. Material was removed in chunks. Often the story line has unexplained gaps. No one seemed to mind.

During the 1952-53 television season, twenty-six Hopalong Cassidy half-hour programs were aired. Fourteen were new, made-for-television Hoppy episodes. The balance consisted of the North Vine films cut down into half hour versions. A voice over narration bridged the gaps in the action. An additional twenty-six new Hopalong Cassidy half hour programs were made for the 1953-54 season.

There is universal agreement among television historians that the quality of the made-for-television Hopalong Cassidy programs left much to be desired. Scripts were weakly written. Action sequences minimal or non-existent. The shooting of outdoor scenes outdoors instead of in studio sound stages lent an air of quality to otherwise drab and dreary looking locations. None of this mattered. Millions of youngsters watched and adored Hoppy.

By the mid-1950s the children who grew up with Hoppy between 1948 and 1954 became young adults. Their interests changed. In addition, a host of cowboy heroes - Gene Autry, Kit Carson, The Cisco Kid, Wild Bill Hickok, and Roy Rogers - arose to challenge Hoppy's television leadership. While some children remained loyal to Hoppy, others switched. When CBS launched Gunsmoke in 1955, the landscape of the television western changed forever. The adult

Western arrived. Hoppy was an early victim, his national syndication ending in 1954. Gene fell in 1956 and Roy in 1957.

In the mid-1990s the tremendous nostalgia kick among the fiftysomething generation has awakened slumbering memories of a handsome elderly man dressed in black astride a white horse. Once again, the Hoppy of the television era rides the range. Will Hoppy influence a whole new generation? No one is certain. However, it is great to welcome back a long lost friend that one has not seen for years.

Learn More: J. Fred MacDonald, *Who Shot the Sheriff?: The Rise and Fall of the Television Western* (New York, Westport, Connecticut, and London: Praeger: 1987).

William Boyd and Clarence E. Mulford

Mulford's opinions regarding Boyd's interpretation of Hoppy appear in numerous sources. Faunce Pendexter in his "Clarence E. Mulford Has Given Accurate Picture of the West" that appeared in the *Lewiston (Maine) Journal* magazine section of April 2, 1949, quotes Mulford as saying that Boyd provides "wholesome, clean pictures. In his portrayal of Hopalong he never commits any acts of which he need be ashamed. Most important of all, he teaches youngsters through the picture that a man can be courageous without being a bully, and that he can have good habits without being a sissy."

Boyd visited Mulford at his Fryeburg home three times during 1949 and 1950. The *Portland Press Herald* of November 25, 1950, indicated that Mulford's "Grandpa stock" with Bruce Perkins, his grandson, "sure went up for having brought to Fryeburg such a famous man as Hopalong Cassidy."

Was it these visits that allowed Mulford to differentiate between Sherman's and Boyd's Hoppy? Perhaps. Or was it this April 15, 1948, letter written by Boyd to Mulford?:

"Dear Clarence:

"Just writing to you makes this probably one of the happiest days of my thirteen years of portraying your, (what shall I call him) let's say your child, Hoppy, and believe me he has been a problem child ever since I first put him into my mind, or into my own skin. I have experienced every feeling you must have had in bringing him into being on paper, plus the fact that I have had to give your word character a body, a voice and a soul. It has been without a doubt, the most interesting assignment I have ever had and I shall always be grateful to you for creating a character that has not only given me a new lease on life, but has given millions of children a lesson in what we both know as, Fair Play.

"It is unfortunate, in a way, that we couldn't have gotten together personally many years ago to blend the feeling and the friendship that I feel sure has developed between us to have brought out an even stronger and better Hopalong.

"I know there must have been times when you wondered what I was trying to do with your Hoppy. As a matter of fact there were times when I wondered myself. The one thing I knew definitely I must do was to keep him as clean as possible, which I succeeded in doing, I think. He may not be the roughest and toughest Western character on the screen but he, at least, has been on the screen longer than any character ever created and that, I am sure, is the most important thing to me. We must also realize that over a period of years we have quite a different audience. I am of the opinion that any child who can speak is intelligent, also that the child who was 12 when I started the pictures, is now 25, so to retain the ones who have grown up and to keep abreast with the intellect of the younger ones, the character of Hoppy has undergone several changes...

"Now I'll get off my soap box for a minute. I do hope you will forgive this outpouring, but they tell me that's what you have friends for, to open up on.

"Let's look at the pictures we have now. Thanks to your clear thinking we are in a position now to cash in on a character that we have both worked hard to keep in the spotlight he should have (the top in his field) with the radio, television and commercial accounts I am sure we will start to cash in on your brain child.

"Clarence, I can't tell you how much I appreciate your prompt action on those contracts. I know it will mean a great deal to both of us. I am going to devote my every ounce of energy I have to the job I have before me to prove my gratitude to you, and also to prove to some of those jerks here that it pays to keep your

nose clean, work hard, and mind your own business, and not go off the deep end simply because we had a few lush years, at the cost of the boys who died in the war to make those lush years possible.

"I am deeply grateful for your generous gift, the books, I shall always cherish them. I do hope I can someday make them into a picture that will do them justice.

"I hope this first letter of mine hasn't upset you, Clarence, as I don't usually run off at the head so much, but I do feel that I am writing to an old friend I haven't seen or talked with for too many years and do hope you will accept this jumble of words and thoughts as coming from an old and very dear friend.
      "Sincerely,
      "William Boyd"

During one of Boyd's Fryeburg visits, Mulford and Boyd both expressed sorrow that neither had meet sooner. While it is fun to speculate on what might have happened to Hoppy if they had, the truth is that it is probably best that they did not. Their visions of Hoppy were poles apart. Each approach is equally valid. How fortunate we are to have both.

Chapter 2

HOPALONG CASSIDY: KING OF THE COWBOY MERCHANDISERS

Roy Rogers may have been Republic's "King of the Cowboys," but he was lesser nobility when it came to licensed merchandise. The "King of the Cowboy Merchandisers" was Hopalong Cassidy. In the three and one-half year period beginning in late 1949 and ending after the 1952 Christmas season Hoppy rode so far ahead of the pack that his competition had to eat his dust.

No matter where one turned, there was Hoppy. He exercised total market control. No single entity, not Barbie, G.I. Joe, the Teenage Age Mutant Ninja Turtles, or Star Wars memorabilia, has enjoyed the market percentage of sales that Hoppy did in the first few years of the 1950s. It is questionable if a single person or character ever will again. It was a Hopalong Cassidy era.

William Boyd and his various business organizations demonstrated to Hollywood and the rest of the world the full extent of the commercial potential inherent in star endorsement. Boyd was a pioneer in product licensing. All the star and promotional projects that followed, from The Beatles to Star Trek, owe a debt of gratitude to William Boyd's vision and performance.

Boyd developed no new marketing techniques. His contribution was linkage, in 1990s terminology, synogism. He showed the connection between product, television, and personal appearance support. He stretched the concept of brand name loyalty to character loyalty. The only recommendation most children needed was a Hoppy endorsement. The openings to two of the Hoppy television trailers say it best:

> "Hi, there....friends. This is the first time in the five years since you've been kind enough to invite me into your homes on television, that I've taken the liberty of speaking to you personally. I would just like to assure you that any product you see Hoppy's name on, in my mind, is the finest...."

> "Hi, there...did you hear what the man said? I did - and I agree with every word of it. If you buy his product I'm sure you'll like it - and it'll tickle my sponsor to death...."

Unfortunately, popularity polls were not as common in the early 1950s as they are in the mid-1990s. Had they been, there is little doubt in my mind that Hopalong Cassidy would have led the list of the most trusted men in America, perhaps far outdistancing the president. The public's confidence in Hoppy was enormous. Boyd deserves credit on two fronts: (a) maintaining that high confidence level through an exemplary lifestyle and (b) utilizing that confidence as a merchandising tool.

Historical Precedents

Boyd was not the first star endorser - far from it. Baseball players, stage stars, other public figures, and cartoon/fictional characters had long been fixtures on a variety of products, e.g., tobacco trading cards. In the late 1930s the Shelby Gum Company, Shelby, Ohio, issued a forty bubble gum card set entitled "Hollywood Screen Stars." Tom Mix and Ken Maynard were among the few cowboy stars featured in the set. A William Boyd/Hopalong Cassidy card was not included. The Hamilton Chewing Gum Company issued the set in Canada in two versions -English backs and French backs. In 1940 Gum, Inc., Philadelphia, Pennsylvania, issued its forty-eight bubble gum "Lone Ranger" card set.

Babe Ruth licensed a manufacturer to make Babe Ruth polo shirts and sweat shirts. In the early 1920s movie greats such as Charlie Chaplin, Douglas Fairbanks, and Mary Pickford supplemented their movie income from royalties received for merchandise licensing. Mary Pickford Skin Cream is an example.

Did Boyd ever receive offers to endorse products, serve as a product spokesperson, or license products in his pre-Hoppy days? The answer is probably no. He was a star, but not of the stature of a Chaplin or Fairbanks. However, Boyd's pre-Hoppy image did appear occasionally, usually on objects promoting a film in which he starred. The easiest items to find are sheet music with Boyd's picture on the cover. "Only for You," the theme song from the P.D.C. Picture *The Leatherneck* is one example. One of Milton Bradley Company's Movie-Land Puzzle sets by A.J. Saxe included an image from *The Yankee Clipper*, one of the DeMille Productions in which Boyd starred.

The leader in merchandising licensing in the 1930s was the Walt Disney organization. Under the leadership of Kay Kamen, Disney characters were licensed throughout the world. Mickey Mouse, Donald Duck, and Pluto became household names. Disney's focus was juvenile products. In analyzing the early products and the age group to which

the Ralston Cereal Company sponsored the Tom Mix radio show. Mix himself never appeared personally on the show. Despite this, thousand of youngsters supported the cowboy who guaranteed "Straight Shooters Always Win."

The owners of the character rights to the Lone Ranger developed the first pre-planned merchandise licensing program. Within a few years, Gene Autry and the owners of the Red Ryder character instituted their own merchandising programs. The field was becoming crowded and highly competitive.

Between 1939 and 1944 Gene Autry created a merchandise licensing base of clothing and toys that put him in a position to claim the title of "King of the Cowboy Merchandisers." Autry's advantage over Hopalong Cassidy was that he was free to exploit merchandising rights since he was not dealing with a copyright character. In fact, while Hopalong Cassidy films continually challenged Gene Autry films for box office supremacy in the late 1930s and early 1940s, Gene won the merchandise licensing war because the owners of the film rights to Hopalong Cassidy did not recognize the potential in merchandise licensing and did virtually nothing.

When Gene Autry entered the service in 1942, his merchandise licensing program lost the key support it needed - a continual supply of new movies. Republic replaced Autry with Roy Rogers who immediately began a merchandise licensing program of his own. By 1946 Roy Rogers' movies outdrew those of Gene Autry and Hopalong Cassidy and his merchandise licensing program exceeded Gene Autry's program. As 1947 drew to a close, the popularity of cowboy stars was: (1) Roy Rogers, (2) Gene Autry, (3) The Lone Ranger, (4) Red Ryder, (5) Cisco Kid, and (6) Hopalong Cassidy.

Hopalong Cassidy Enters

Harry "Pop" Sherman, owner of the movie rights to Hopalong Cassidy, and William Boyd had parted company early in 1944. Sherman wanted to produce "A", not "B" movies. Boyd survived touring with various circuses - Hoppy's untapped potential weighing heavily on his mind. Boyd believed in Hoppy. Sherman's limited vision held Hoppy's full potential in check. If only Boyd could find the financial backing to make some additional Hoppy movies, he would shape the character to his vision and show the film industry that he was correct.

The man on the white horse in this instance was Toby Anguish. Anguish put together a group of financial backers to make an additional series of six Hopalong Cassidy films. He asked Boyd to contact Sherman and negotiate the rights for the Hopalong Cassidy character. Sherman, questioning Boyd's seriousness, offered Boyd's group not only the movie rights to the Hopalong Cassidy character, but the theatrical re-release rights to forty-eight of the fifty-four Hopalong Cassidy movies that Sherman had produced as well. The price was $250,000. To Sherman's amazement, Boyd agreed. William Boyd Productions was born in 1946.

Toby Anguish sold the theatrical re-release rights to Film Classics, an independent distributor, and negotiated a deal with United Artists to release the new series. William Boyd Productions had the funds to proceed.

Although William Boyd is listed as executive producer in the credits, Lewis J. Rachmill was the on-site producer of the twelve North Vine Hopalong Cassidy feature films. The

they would most appeal, the emphasis was on ages two to ten. The Disney concept also was broad. Every aspect of a person's life from the toy with which they played to the food that they ate was fair game for a Disney license. In the early 1930s Disney teamed up with the Bureau of Public Relations of the American Dental Association to issue a one and one-quarter inch, celluloid, pinback button featuring Mickey Mouse defeating the Big Bad Wolf with a toothbrush, the slogan surrounding the picture reading "MICKEY MOUSE / GOOD TEETH."

The Disney licensing techniques served as a model for Boyd and others. None followed them as closely as Boyd. However, Boyd made one significant change. He shifted the target audience from the two to ten age group to the seven to fourteen age group for his products. Boyd totally ignored the infant market. He wanted a vocal market, a market where his followers could verbally tell their parents, "I want one of those."

Early Cowboy Merchandise Licensers

Likewise, the concept of western character merchandise licensing was well established before Boyd as Hoppy rode the range. Tom Mix, the first major western star in sound motion pictures, signed a number of merchandise licensing agreements with manufacturers. Mix took opportunities as he found them. He had no formal licensing program. Mix is best known for premiums associated with the Ralston Straight Shooters, a promotional concept of the Ralston Cereal Company. From 1933 until the early 1950s,

Hoppy team consisted of William Boyd as Hopalong Cassidy, Andy Clyde as California Carlson, and Rand Brooks as Lucky Jenkens. *Devil's Playground*, the first of the North Vine films, was released on November 15, 1946, a little less than two and one-half years after *Forty Thieves*, Sherman's last Hoppy film, was released. *Strange Gamble*, the last film in the series, was released August 8, 1948.

While Boyd's Hoppy films did not lose money at the box office, they did not produce the profit levels for which United Artists and Boyd himself had hoped. Boyd wanted to produce a third series of six films. United Artists expressed no interest in distributing them. The decision appeared to mark the end of the trail for Boyd and Hoppy. In reality, it was only the end of the Chapter Two.

Hoppy Moves To Television

1948 marked the halfway point of the filming and release of the final twelve feature-length Hopalong Cassidy films, known to Hoppy collectors as the North Vines. Boyd's career was at a critical point. He was typecast as Hoppy. United Artists, the distributor of the North Vine films, had no interest in financing a third series of six Hoppy titles.

Boyd in his early fifties was too young to consider retiring. Like his fictional counterpart, it was time to tighten his belt, move on, and seek the challenges of a new frontier. To Boyd's credit, he did.

Initially the film industry fought television. The major studios prohibited their stars from appearing on television programs and refused to release films from the archives for showing. Foreign and low budget films were the principal fare. An opportunity awaited an independent film producer ready to take a risk. Boyd and Anguish recognized the opportunity and took it.

Television rights were included in the rights purchased from Sherman. Toby Anguish raised $50,000 to make new 16mm prints of the films for TV distribution. In addition, in an agreement dated June 21, 1949, Toby Anguish purchased the rights to the first six films that Sherman had sold earlier from Goodwill Pictures, Inc.

Doubleday, publisher of the Mulford Hoppy novels, acting on behalf of Clarence E. Mulford, raised the question of whether or not Mulford had actually sold the television rights to Sherman. Caught between the proverbial rock and hard place and facing a lengthy and costly legal battle, Boyd settled with Doubleday and Mulford plus purchased a quitclaim from Paramount Pictures, the studio that had released the early Hopalong Cassidy films. The moneys required for the additional rights purchases virtually exhausted all Anguish's and Boyd's available funds. Their *strange gamble* would paid off handsomely.

Boyd and Anguish offered the Hopalong Cassidy films to television. There was a stampede. Hopalong Cassidy films began appearing on a regular weekly basis in the New York market in November 1948. NBC quickly struck a deal to feature the films as part of its prime time network programming. Hoppymania was born.

From 1949 until the mid-1950s Hoppy rode roughshod over the television range. His success was phenomenal. In 1953 NBC Film Production produced a promotional booklet entitled "Hopalong Cassidy outdraws 'em all!" The marketing figures are staggering for a cowboy riding the range for a fourth season. Hoppy's ratings in the Boston market were six times greater than the Lone Ranger's. During his time slot in the New York market, he had a 66% audience share. In Cleveland his audience share reached 79%. He even outdrew Howdy Doody. In Chicago the only program he could not outgun was White Sox baseball.

The most impressive statistic of all is that 46% of the Hopalong Cassidy audience was adults. Hopalong Cassidy's high approval rating with adults, especially those with young children, is not well understood by collectors. Further, he appealed equally well to both sexes. Young girls and mothers were among Hoppy's biggest supporters.

Hoppy's Public Persona

The relationship between Harry Sherman and William Boyd was one that necessity dictated. At its best it was tolerable, at its worst strained. They needed each other and grudgingly accepted the fact. Most of what is known is hearsay. The principles are dead. Grace Bradley Boyd refuses to discuss the relationship on the ground that "if you cannot say something nice about somebody, etc., etc."

What is known is that Boyd did little to promote Hoppy during the Sherman film period. Why? Was it Boyd's reluctance to promote a character in which he had no financial interest or Sherman's failure to see the value in promoting "B" movies? The answer most likely rests somewhere in the middle.

Boyd's Hoppy promotional activities during the North Vine period also were limited. Limited finances had to be a factor. All available moneys were tied up in rights purchases and production. In addition, the North Vine films competed head-to-head in the theaters with the re-releases of the Sherman Hoppy films.

There is little question that Boyd recognized by the early 1940s a clear connection between public appearance exposure and the successful exploitation of the Hopalong Cassidy character. Numerous 1940s references to this fact are found in the Boyd papers and interviews. Given the chance, Boyd would prove he was right. Television provided that chance.

This is not the place for a chicken and egg argument. Did Boyd's willingness to commit himself to an exhaustive schedule of personal appearances or the appearance in stores of hundreds of licensed products lead to the Hoppymania of the early 1950s? Each drove the other. The merchandising effort would have been much less successful had Boyd not supported it through public and media appearances. The appearances would have been for naught had there not been an unending supply of Hoppy products to purchase. Discussing Boyd's personal appearance schedule first is not putting the cart in front of the horse.

Boyd took advantage of any and every opportunity offered to him. He appeared as a guest on numerous radio shows, mostly of the variety type. He worked with, not against, the press, appearing on a host of magazine covers in 1950. He shook hands at dozens of department stores across the country. The Carousel Parade in Charlotte, the Mardi Gras Parade in New Orleans, the Rose Bowl Parade in Pasadena, and the Thanksgiving Day Parade in Philadelphia are just a few of the parades in which he appeared.

Boyd realized that a broad support base was essential to fuel the public's passion for Hoppy. Movies and television were givens. Time to cover the other bases as well. Boyd proceeded to launch a Hopalong Cassidy radio show. A

Hopalong Cassidy comic strip appeared in over 150 newspapers. Three million Hopalong Cassidy comic books sold every month. All kept his image before the public. All provided a vehicle to advertise and sell Hoppy merchandise.

Boyd created a vertical merchandising empire. Others had attempted it previously. Some met with limited success. None had Boyd's vision, drive, or persistence. The "Boyd Model" remains the standard today. To use a line from a James Bond movie, "No one did it better."

Hoppy Merchandise

The keystone to the Hoppy empire was merchandise licensing. $20,000,000 worth of licensed Hopalong Cassidy and William Boyd products were sold in 1952. It was not a record year.

The first Hopalong Cassidy merchandise license was signed on April 29, 1949, between Hopalong Cassidy, Inc., one of many business entities that Boyd would create between 1948 and 1972, and Eli E. Albert of 935 Broadway, New York, New York, for the manufacture of jodhpurs, slipcovers, and jackets. Two days later, May 1, 1949, Yale Belt Corporation became licensee for children's leather belts, suspenders, and leathercraft kits. Four additional licensing agreements would be signed in 1949: Goodyear Rubber Company, Middletown, Connecticut, October 4, 1949, for rubber cowboy boots; D.P. Harris Hardware Manufacturing Company, New York, New York, December 14, 1949, for bicycles and roller skates; Morse and Morse, Inc., Los Angeles, California, April 11, 1949, for T-shirts, polo shirts, and pajamas; and, Oxford Boyswear, Inc., New York, New York, June 28, 1949, for boy's dress pants and slacks. An empire had begun.

1950, 1951, and 1952 were banner years for Hopalong Cassidy merchandise licensing. Before the smoke cleared, over 2,500 products would be marketed bearing the image of Hopalong Cassidy or William Boyd. Some were in series, most stood alone. The number of Hoppy items increases significantly when commercial endorsements, e.g., the use of Hoppy's image in a magazine advertisement, and industry premiums, e.g., a paper pop pistol used by many different businesses, are added to the list. Collectors have uncovered examples of all known licensed products that were marketed. However, new finds of endorsement and premium material still occur. This is what makes collecting Hoppy so challenging.

No one who grew up during the early 1950s would dispute the claim that Hopalong Cassidy was "King of the Cowboy Merchandisers." Hoppy merchandise was everywhere - from the brush used to comb one's hair in the morning to the bed sheets and covers into which one crawled at night. If you need proof, check Appendix #1, the list of Hopalong Cassidy licensees.

The introductions to each section with "Hopalong Cassidy Merchandise," this book's fifth chapter, provide general market information about specific licensed product groups. Repetition accomplishes little. Instead, I have chosen to focus here on the licensing contract itself, the individuals in the Hopalong Cassidy merchandising effort, two case studies as typical examples, and a general look at the dairy industry.

However, a few general observations before getting to the nitty gritty. The vast majority of Hopalong Cassidy merchandise was seasonal. The bulk of sales for most companies occurred between August and December, the traditional months when stores stock and sell Christmas merchandise. There is little question that Hoppy merchandise was gift driven, with Christmas followed by birthdays as the principal reason for purchases. The other major selling period was August when parents bought Hoppy licensed products for youngsters heading back to school.

Analysis of hundreds of royalty statements also suggests that most products had a maximum sales life of two seasons. Three or more seasons was exceptional. U.S. Time's Hopalong Cassidy wrist watches and Aladdin's lunch kits and lamps defied the odds and remained in production for over five years. The peak period for Hopalong Cassidy merchandise was the 1950, 1951, and 1952 Christmas seasons.

It is important to remember that the scope of the Boyd licensing program was pioneering in nature. Track records were minimal in respect to how long a marketing phenomenon such as Hoppymania would last. Tragically, many Hoppy licensees thought the good times would never end. Based on their initial successes, they signed contract extensions, all of which required non-recoverable up front advances against royalties and a guaranteed annual income. As early as the first quarter of 1953 manufacturers were writing to Boyd asking for relief. To his credit, it usually was granted.

Boyd's flexibility is admirable, but expected. He and his advisors knew that they were blazing new territory. They used previous models, but revised them to fit their needs. They quickly learned about cutthroat competition. In order to cut into Hoppy's market, Roy Rogers and Gene Autry agreed to licensing royalty percentages significantly below that requested by Hoppy. The business frontier of the early 1950s was as wild as the wild west of the 1870s and 1880s. Hoppy felt right at home.

Learn More: *Life*, June 12, 1950, pages 63-70; *Time*, November 27, 1950, pages 18-20. Far too many individuals who write about Hoppymania paraphrase or quote directly the information that appears in these two articles. I recommend that you read them for yourself. Most large metropolitan libraries have *Life* and *Time* on microfilm or have access to a service where photocopies can be obtained. It is for this latter reason that I included the page numbers for these two articles. Do not forget to request to have the cover copied. It will serve to document the source. In addition, it features Hoppy.

One of my reasons for including the *Time* article is that it is one of the few early 1950s articles that has negative comments about Hoppymania. It is worth reading as a balance to all the positive information that appears, including the overall tenor of this book.

The Merchandise Contract

The typical merchandise licensing contract for "the right to use the name 'HOPALONG CASSIDY' and the likeness of William Boyd as HOPALONG CASSIDY in connection with the articles specified in Paragraph 22 hereof" was ten, single spaced, type-written pages in length and contained twenty-three sections. Most merchandise licenses were granted for a two or three year period. If the total royalties fell below $10,000.00 per year, Hopalong Cassidy, Inc., reserved the right to give thirty days notice and terminate the contract.

Boyd and his representatives were fanatical about quality. Section 3 began "You agree that within a period of ____ (normally 60) days from the date hereof you will make up samples of each and all of the articles specified in Paragraph 22 hereof at your own cost and expense, and submit them to us for our written approval..." Later in the same paragraph one finds "We, naturally, shall have the right to reject the samples submitted by you, whether because of the quality of design of said samples or otherwise..."

Boyd and his group rejected samples that did not come up to their standards. In addition, they provided for the regular receipt of production line samples so that the quality level that they demanded would be maintained. Boyd also had packaging approval. Section six contains the following: "The packaging of the approved articles must conform to the samples thereof which have been submitted and approved by us."

Initially the product royalty was five percent of the gross receipts from sales. Boyd stood firm behind this figure between 1949 and 1952. As competition developed from other television cowboy heroes and Hoppy's popularity faded somewhat, the figure was occasionally reduced. Rarely did the rate drop below 3%. A non-returnable advance against royalties, usually ranging between $2,500 and $5,000, was required upon signing a contract.

Boyd was very specific about exactly what he was licensing. Milton Bradley, Springfield, Massachusetts, received a license for jigsaw puzzles, inlay puzzles, and regular children's puzzles. Western Printing & Lithography Co., Racine, Wisconsin, also received a license for jigsaw puzzles. At first glance, it would appear that Boyd licensed two manufacturers to produce the same product. Not true. Milton Bradley's license was restricted to "jigsaw puzzles, inlay puzzles, and regular children's puzzles (similar to jigsaw puzzles except the pieces do not interlock) in such design as will be approved by me, to retail at not less than 50¢." (License dated February 10, 1950, between Hopalong Cassidy Enterprises and Milton Bradley.)

Price level was just one of the means Boyd used to create licensing parameters between manufacturing of similar products. Quality was another. Boyd knew that if he was to be successful, he had to license products with broad financial appeal. However, he resolved never to sacrifice quality for profit. Boyd insisted on the best quality possible for the price. This was critical if he was going to retain the confidence and trust of parents and children.

Boyd carefully limited the territorial boundaries in his licensing agreement. Most manufacturers were licensed to sell products only with the United States. Canadian, Central and South American, English, Continental, and Far East licenses were negotiated separately. English agents handled most of the British Commonwealth licensing. The Hoppy foreign markets developed two to four years behind the American market. The Boyd business papers contain numerous letters from several American manufacturers who saw marketing opportunities abroad. Today what was obvious to so few would be take for granted by all. As stated earlier, Hoppymania was a pioneering effort. Much was tried. Hoppy led the way.

During the 1950s and right up to his death, Boyd continually restructured not only his licensing, but his other business activities as well, as new opportunities presented themselves. Business names changed continuously. Hopalong Cassidy Enterprises joined Hopalong Cassidy, Inc. There was a Hopalong Cassidy Company. The key is that none of these changes marked a radical change in the manner by which Boyd did business. Business as usual was a Hoppy trademark.

The Exclusivity Clause

One clause in the early Boyd licensing contracts came to haunt Boyd and the manufacturers who signed with him. Section 9 read: "You further agree that you will not manufacture, sell or distribute any of the articles specified in Article 22 hereof or any article directly similar thereto which is identified with or bears the name of any living or dead person of prominence or any cowboy personality."

Simply put, if you did business with Hoppy, you did not do business with any other fictional or real cowboy, living or dead. When Hoppy's popularity skyrocketed and manufacturers were clamoring to jump aboard the Hoppy bandwagon, their judgment was clouded. Few thought carefully about the long term implications of this clause. No one can fault Boyd for asking for it.

A memorandum dated March 27, 1953, from Dan Grayson of HC Promotions Association to Marguerite Cherry concerning Aladdin Industries, licensee for Hopalong Cassidy lunch boxes, reads: "Sears Roebuck placed Roy Rogers with the Thermos Company. As you know, this is number one in the vacuum bottle and lunch kit field and they are Aladdin's biggest competitor. From information we have received from Aladdin and the trade we understand that the royalty rate that Rogers contracted for with Thermos is 2 ½%. Thermos came out with the Rogers lunch kit at five cents less per unit at the wholesale and chain store levels. This was quite a blow to Aladdin as Hoppy no longer has the strength to fight a five cent per unit differential. At best it has become an even-steven fight if everything is taken into consideration. The slight edge that Hoppy might have on Rogers' popularity is off-set by the amount of Rogers' goods purchased by Sear's Roebuck..." Aladdin wanted relief. Boyd responded by increasing Aladdin's foreign market rights.

A letter from V. K. Church, Vice President, of Aladdin Industries, Inc., to Hopalong Cassidy Co., dated November 2, 1954, reads "While the final figures are not tabulated as yet, country-wide, I believe we will have sold many more Hoppy kits by the end of this year than we did last. However, on the West Coast, where Rogers apparently has the kids in his pocket, this will not be true. The thing which helped the sales of Hoppy kits most this year was the redesigning and re-styling of the kit and bottle which we feel make our kit stand head and shoulders above the others..." Aladdin was still loyal to Hoppy, but the loyalty would last less than one month.

On December 2, 1954, V. K. Church once again wrote to Hopalong Cassidy Inc.:

"When Aladdin introduced Hopalong Cassidy school kits to the trade in May 1950, it was the first time character merchandising had ever been used in the vacuum ware industry. Since then, well over a million Hoppy kits have been sold and the resultant royalties are approaching the $100,000 mark. They are actually $94,652.65 when I check(ed) last.

"As was only to be expected, eventually, our competitors recognized the value of using well known characters in connection with the marketing of school lunch kits; and today, a child can choose between a Hoppy, Gene Autry, or Roy Rogers kit - to name those personalities currently being used by manufacturers in the vacuum ware industry.

"Last year, Rootie Kazootie, Howdy Doody, Superman and Lone Ranger school kits were introduced by yet another manufacturer, who, while unable to furnish vacuum bottles with these kits, never-the-less sold these kits in substantial quantities...

"To meet this competition and hold our leadership in the field, it is going to be necessary for us to broaden our line. To do this, we are adding a Wild Bill Hickok and an Annie Oakley kit to the line..."

Aladdin had no choice and neither did the Hoppy interests. The exclusivity clause was amended. Eventually it would disappear entirely. To Aladdin's dismay and Thermos' delight, Boyd's previous insistence on exclusivity gave rise to a competitor who had greater flexibility and was able to overtake Aladdin as the leader in the vacuum bottle field. The story probably would have ended differently had Aladdin been able to license other character lunch kit as market interest shifted.

Knockoffs and Shady Characters

Knockoffs were as much a problem in the 1950s as they are in the mid-1990s. There were direct copies and Hoppy look-alike copies. The Hopalong Cassidy copyright did not grant Mulford, Boyd, or others the exclusive rights to an image of a cowboy dressed in black sitting astride a white horse. Each year I see several supposedly rare Hopalong Cassidy licensed items that are from this knockoff group. How do I know? I know what products Boyd licensed and which he did not. Study and memorize Appendix #1, and you will know as well.

When a manufacturer no longer wanted to continue a license, Boyd's staff tried to find a replacement. More often than not, there was someone waiting in the wings. However, the transition was not always smooth. There were shady characters in the wild west as well as business community of the 1950s.

On October 29, 1951, Robert P. Cable, president of Cable Raincoat Company, sent a letter to Hopalong Cassidy, Inc., that read:

"Our salesmen have been reporting to me that they have seen boy's raincoats with the Hopalong Cassidy insignia of material and style similar to ours at various stores and that these coats were not of our manufacture. One of our men was in a store in Hartford, Con-necticut when a shipment of such raincoats from Spatz Brothers was received. Another one of our men saw it in one of the large department stores in New York City.

"This is contrary to our understanding and I would appreciate hearing from you at once with reference to the same."

On November 12, 1951, Marguerite Cherry responded to Cable's letter indicating that an investigation had been launched and requesting any additional information Cable had uncovered. An inter-office memo of the same date from Marguerite Cherry to Dan Grayson asked him to check if the Spatz shipment related to a back order. A Spatz raincoat license had expired on December 31, 1950. Unfortunately, Boyd's group forgot to include a provision limiting the time period during which existing stock could be sold once a license had expired.

Cable wrote a second time on December 7, 1952, about the problem. On January 31, 1952, he wrote again: "We have not heard from you further in connection with this matter and feel the situation merits your attention." Cherry responded on February 6, 1952, indicating that Boyd's attorneys, Gang, Kopp & Tyre were handling the matter and had written to Spatz Brothers and the two stores trying to solve the problem. Lee Spatz of Spatz Brothers, New York, New York, responded on February 7, 1952 indicating that they had "not sold any Hopalong Cassidy products since the termination of our contract."

On May 15, 1952, a letter from Gang, Kopp & Tyre was sent to the Cable Raincoat Company that made three points: (1) "our client is no guarantor that there will not be infringers or persons unauthorized by it who are imitating or selling Hopalong Cassidy products," (2) Boyd and his organization was doing everything possible to "track down any sales of unauthorized Hopalong Cassidy raincoats," and (3) in response to Cable's claim that the contract had been breached without justification, Hopalong Cassidy, Inc., is demanding termination and full payment of the guarantee of $25,000.

It is common knowledge that once lawyers become involved in a problem, a simple solution is impossible. The matter continued to drag out through the end of 1952 and into 1953. By May 1953, the matter was before the American Arbitration Association. Cable produced ample evidence to suggest that Spatz had lied to Cherry and other Boyd representatives. A compromise was reached; and, the Cable raincoat license terminated.

Both parties were losers. Cable made and sold no raincoats during the controversy. They received no income; and, hence, Boyd received no royalties. Both incurred legal expenses.

Could the situation have been handled better? Absolutely. However, licensing, infringements, and the difficulties of keeping hundreds of manufacturers happy was a relatively new experience to Boyd and his team. Given this, it is surprising that they did as well as they did.

The Triumvirate

Who ran the Boyd licensing empire? In terms of setting policy and determining a course of action, the answer is William Boyd. In terms of the day to day logistics of a licensing business, the answer was a triumvirate - Marguerite Cherry, Dan Grayson, and Lewis E. Pennish.

Robert Stabler, Boyd's business and road manager and later Vice President of Hopalong Cassidy, Inc., worked closely with the triumvirate. Because his activities often required him to be away, he is not included as a member of the group. A case can be made either way.

Marguerite Cherry was the key player. Her official title was Secretary-Treasurer. While I found no record of any unofficial title, I cannot help thinking about her as the "Mother Protector." She not only ran the office, she protected Boyd and the Hoppy image from anything that might stain them. Her loyalty was total, lasting until her death. I do not know what Hoppy would have done without her.

Grayson headed Grayson & Associates, Inc., a licensing agent and merchandiser, located in Beverly Hills, California. He recruited potential licensees and did the follow-up. When Cherry had a question or problem, she turned it over to Grayson. He made the necessary contacts, gathered the need information, and reported back with a recommendation.

Lewis E. Pennish was located in Chicago and provided services similar to those of Grayson. He apparently traveled to New York City frequently. The business files contain frequent memos regarding his meeting with clients in New York. His correspondence suggests that he also was active in the public relations sector.

Do not be fooled by this brief mention of these individuals. They were the Hoppy team and their roles were critical. Because they performed their duties well, Boyd could devote his full energies to the public promotion of Hoppy. He did so knowing he left the ranch in good hands.

Case Studies

Grace Bradley Boyd, William Boyd's widow, placed William Boyd's personal and business papers in the public domain. She deserves the highest praise for this foresight and consideration. Access to Boyd's licensing contracts, correspondence, and royalty statements provides a wealth of insights to merchandise licensing and promotion in the 1950s.

Space does not permit a detailed licensee by licensee analysis. This will have to wait for a scholarly researcher. For now, the following two case studies will have to suffice.

No. 1 - Milton Bradley

Milton Bradley manufactured nine different products under its Hopalong Cassidy merchandising license, two of which were issued in an updated version. All are familiar to Hopalong Cassidy collectors. Production ranged over seven years, 1950 to 1956. The vast majority of production was centered in the years 1950 to 1953.

Perhaps the best known Milton Bradley product is the Hopalong Cassidy Game, No. 4047. Between 1950 and 1953 Milton Bradley sold 369,054 units of the first version of the game. A second version, No. 4047A, issued in 1953, remained in production until 1955. 52,188 units were sold. The total sales of the two versions is 421,242 units. Milton Bradley's most successful item was its No. 4025 Hopalong Cassidy puzzle. The first version sold 578,486 units between 1950 and 1952. A second version, 4025A, was made between 1953 and 1955 and sold 57,352 units. Total sales for the two versions was 635,838 units.

In 1951 Milton Bradley launched its Hopalong Cassidy Television Puzzle set. The principal sales period was 1951 to 1954 when 60,356 units were distributed to stores. 1955 sales were limited to 36 units. 1951 also saw the initial distribution of Hopalong Cassidy Dominoes, No. 4104. 66,549 units were sold in three years, 1951 to 1953.

The Hoppy Village, No. 4108, issued in 1951, sold only 30,008 units. Its annual sales indicate a typical sales pattern; 1951 - 17,305 units, 1952 - 6,320 units, 1953 - 12,665 units, and 1954 - 5,117. Never once did sales the second year equal or exceed those of the first year. A drop off in sales of fifty percent was common.

Milton Bradley's last new product was its Hopalong Cassidy Television Inlay Puzzle. No. 4508-15. During its initial year, 1955, the puzzle sold 13,482 units. Sales in 1956 amounted to 7,486 units. The puzzle was the only Hopalong Cassidy item offered by Milton Bradley in 1956.

Initially Milton Bradley was licensed only to produce jigsaw puzzles that retailed for over fifty cents. A Memorandum dated November 5, 1952, from Marguerite Cherry to Dan Grayson, reads:

"Please amend Paragraph 22 (2) so that subject company can make jigsaw, inlay puzzles, and regular puzzles, etc., to retail for less than 50¢. The Western Printing contract has expired and has not been renewed.

"Bradley is doing an excellent job in games of all kinds and is in a position to expand their volume with the low range of puzzles. I just spoke to James Shea, the President of this company, by phone and he is enthused with the way that Hopalong sales have held up."

As already indicated, the Milton Bradley sales picture changed. A letter dated May 18, 1954, from Millens W. Taft, Jr., Director, Research and Development for Milton Bradley, reads:

"This is to inform you that as of our royalty period ending March 31, 1954, the following two items have been sold and discontinued from our line: 1) 4104 Hopalong Cassidy Dominoes. 2) 4108 Hopalong Cassidy Frontier Village...

"We have enjoyed very much our association with Hopalong Cassidy Company, but due to poor sales on the above two items we felt it was wise to drop them from our line. However, the name Hopalong Cassidy is still an important one with us, and we trust that our relationship will continue for many years to come."

A Taft letter dated April 26, 1955, indicates that 4103 (Hopalong Cassidy Television Puzzle), 4105 (Hopalong Cassidy Puzzle), and 4308 (Hopalong Cassidy Western Inlaid, will be replaced by 4508-15 Hopalong Cassidy Western Inlaid). By 1956 discussions were initiated between Taft and Dan Grayson regarding acquisition of Milton Bradley's plates, art work, and dyes by the Hoppy interests. Hoppy would no longer ride the Milton Bradley range.

No. 2 - U.S. Time

The United States Time Corporation received a license from Hopalong Cassidy, Inc. to manufacture: "Wristwatches, pocket watches of every nature and description. Also alarm clocks, all in such style and design as will be approved by us." An important exception to the exclusivity clause was granted: "We acknowledge that you manufacturer for distributors (but do not sell) the Lone Ranger watches. You may continue to manufacture such watches but not exceeding 200,000 watches during any year of the term..."

The U.S. Time license was one of the most successful licenses issued by Boyd. The first Hopalong Cassidy wrist watches were sold in April 1950. Between April 1950 and January 1953, U.S. Time sold 841,399 units. In August 1950, U.S. Time introduced their Hopalong Cassidy alarm clock. It sold 38,372 units by January 1953. While the sales of U.S. Time's wristwatches and alarm clock were satisfactory, the sales of their Hopalong Cassidy pocket watch left much to be desired. Only, 7,437 units sold between February 1951 and September 1952, at which time is was removed from inventory.

An individual who owned a U.S. Time character watch in the early 1950s can appreciate the following letter sent by William Boyd to Lloyd Bernader of the United States Time Corporation on June 18, 1951:

"I am writing to you about a matter which is of the greatest concern to me. For some time now you have been manufacturing the Hopalong Cassidy watch. You may recall that it was with some trepidation that the manufacture of a watch was undertaken because I wished to be assured that the watch would be of high quality. I could conceive of nothing worse than selling hundreds of thousands of watches which were of poor quality and which created dissatisfaction in the minds of consumers.

"For some time now we have been getting complaints in snowballing quantities. I am enclosing copies of two of these complaints, which have already been forwarded to you from my office, which are only typical of the many we have received.

"This last Christmas I had occasion myself to distribute a number of Hopalong Cassidy watches as Christmas presents. I have checked with a number of recipients of these watches to find out whether their children were able to use the watches for any length of time. The replies to my inquiries were negative. One complaint was that the watch required winding several times a day. Another complains was that the watches were very difficult to wind. Finally there were complaints of the nature of the enclosed letters; the watches stopped, or the stems would fall out.

"You must remember that the Hopalong Cassidy watch is sold for use by children. It is therefore no answer to criticisms of the watches that in the hands of your factory technicians these watches will function properly. The Hopalong Cassidy watch when sold to the public must be a sturdy, rugged, easy-winding watch which will stand a reasonable amount of abuse from children...."

A reply was sent to Boyd on June 27, 1951. A complaint from Hoppy deserved immediate attention. The usual excuses are offered, e.g., what is happening to us is happening to other manufacturers as well. The problem with the stems is admitted. Boyd was assured that it "was give immediate attention and was rectified."

Apparently the problems were resolved to Boyd's satisfaction. Hoppy's relationship with U.S. Time continued into the late 1970s. When the smoke cleared, the sale of U.S. Time's Hopalong Cassidy watches numbered in the millions. This is why they are so easy to find today.

Food Products

Beginning in 1950, Hopalong Cassidy licensing was extended to the food industry, a lesson learned from successful food merchandise licensing efforts by Walt Disney Enterprises. Three food licenses were issued in 1950: Bol Manufacturing Co., Chicago, Illinois, August 25, 1950, for fruit concentrates in all flavors; Kuehmann Foods, Inc., Toledo, Ohio, July 10, 1950, for potato chips; and, Topps Chewing Gum Co., Inc., Brooklyn, New York, July 1, 1950, for hard candy, taffy, and chewing gum. Collectors will immediately identify with the Hopalong Cassidy potato chip cans and non-sport trading cards that resulted from the Kuehmann and Topps licenses. The Van Camp Sea Foods, Inc., Terminal Island, California, license of June 17, 1952, linked Hoppy with Chicken-of-the-Sea tuna.

The bakery industry was targeted from the beginning of the food licensing effort. On July 1, 1950, a license was issued to the Burry Biscuit Corporation. In 1951 members of the Quality Bakers of America Cooperative were licensed to use the Hopalong Cassidy image. Hoppy's endorsement of Bond Bread made it a market leader in its communities.

However, Boyd's and Hoppy's greatest impact in the food industry occurred within the dairy industry. Many youngsters in the early 1950s drank milk or ate ice cream because Hoppy's picture was on the bottle or package. The first dairy license was issued on August 21, 1950, to the Luick Ice Cream Company of Milwaukee, Wisconsin, to manufacture Hopalong Cassidy ice cream. The first milk/ice cream license was issued to Marin Dairyman's Milk Co., San Francisco, California, on April 16, 1951, and covered the counties of San Francisco, Santa Clara, San Mateo, Santa Cruz, and Marin. Numerous other West Coast dairies quickly joined the bandwagon. By 1954 Hopalong Cassidy-endorsed milk products could be found nationwide.

Hoppy World Wide

Boyd's licensing of Hopalong Cassidy extended worldwide. It is another aspect little understood by American Hopalong Cassidy collectors. Appendix #1 contains information on Canadian and other foreign licensees. Their number is significant. Hoppy's influence was not isolated. Children from Europe to the Far East knew who Hoppy was.

Note that the list of foreign licenses includes a number of American firms. The early 1950s provided a unique opportunity for many American businesses. The vicissitudes of war changed trade relationships between America and both its Eastern and Western trading partners. Based on modern practices, the Hoppy licensing list is small, a mere drop in the bucket. However, it was a starting point. Others would build on it. It showed the way, just as numerous other Hoppy innovations did as well.

Hopefully, this book will reach Hopalong Cassidy collectors abroad. I encourage them to write about Hoppy licensing in their countries and publish it in the American trade papers. I know that any well researched and written article will find a home in *Antique Week* (PO Box 90, Knightstown, IN 46148). Of course, I also would welcome learning more about foreign Hoppy licensed materials. Write me at: 5093 Vera Cruz Road, Emmaus, PA 18049.

Hoppy's Influence

Between 1949 and 1952 Hopalong Cassidy was the leading cowboy hero on television. He held his own against his challengers between 1953 and 1955. His influence would continue for another decade, especially in foreign countries. No star has been able to sustain merchandise licensing vitality for such an extended period since.

How influential was Hoppy in the 1950s? Cecil B. DeMille, when interviewed about Hopalong Cassidy for a December 1950 *Coronet* article, said: "Every kid needs a hero. Hopalong Cassidy takes the place of Buffalo Bill, Babe Ruth, Lindy and all the rest. He's everything that young America admires and wants." An article in *The American Weekly* noted: "Today more children are influenced by Hoppy's code than any other single factor in America."

Hoppy was all that was the best in America. Boyd actually became the fictional character that he created and lived his life accordingly. For millions of boys and girls, Hopalong Cassidy was America's cowboy hero. He certainly was mine.

Chapter 3

THE GRACE BRADLEY BOYD INTERVIEWS

On May 24 and May 26, 1988, I had the privilege of recording two lengthy tape interviews with Grace Bradley Boyd, "Mrs. Hopalong Cassidy." Grace set no restrictions on what I could ask, willingly and graciously answering every question. She shared with me information covering a broad range of topics, ranging from the filming of the Hopalong Cassidy movies and television programs to personal reflections on what it was like to be married to the nation's No. 1 cowboy hero of the early 1950s.

The transcripts of our conversations consume forty-three, single-spaced, typed pages. I consider these interviews among the most valuable Hopalong Cassidy research materials in my possession. I am honored to share some of Grace's recollections and insights.

The presentation is organized into six topic areas: Grace Bradley Meets Bill Boyd; Topper; On Location; Merchandising Hoppy; Hoppy Cares; and, A Final Note. I have taken the liberty to edit the material to make it more readable and did a little cutting and pasting to make it flow better. I have not altered or in any way changed Grace's thoughts.

The interviews confirm several stories that have circulated verbally among collectors. In addition, there is a wealth of new information about William Boyd, the role he played in shaping and molding the Hopalong Cassidy character, and how Boyd's adoption of the Hoppy persona effected his personal life.

GBB = Grace Bradley Boyd
HLR = Harry L. Rinker

Grace Bradley Meets William Boyd

HLR: How did you meet William Boyd?

GBB: I have to preface my story by saying that I fell in love with him when I was twelve years old. I had just entered high school. I had his name all over my books - on the outside edges. I had pictures of him that I cut out. I had never seen him in a movie at that point. I feel in love with Bill from his pictures.

After I arrived in Hollywood, I saw Bill a couple of times, but he did not notice me. At the time, he was three weeks away from finalizing a divorce. An agent, a mutual friend, told him, "Grace. Grace Bradley. Now, there's a girl you should know. She's a heck of a gal."

It stuck in his head. Shortly after, he was driving down to his beach house when he stopped at a light. The agent pulled up beside him. Bill called over to the agent and asked him to call Bill at the beach house with my phone number.

He called me. He said he would like to meet me. Everyone who knew me, knew how crazy I was for him. I had been carrying a torch for all these years. I thought someone was pulling my leg, so I was very cool. When I realized it was him, I was absolutely paralyzed, but still able to respond when he asked me for a date.

We made the date for Saturday at the beach house. He said I would be well chaperoned. There were three other couples coming to the beach house. At the time, I was living with my mother in a little town house in Beverly Hills. He would pick me up there.

He arrived with a friend. My mother opened the door. Bill said later that for a second he did not know if the person answering the door was my mother or me. My mother told him, "Grace will be down in a minute." Then he knew.

I made my entrance down a stair case. Bill tells it, "I saw the feet. I saw the legs. I saw the body. I saw the face." He put his arms out; and, I walked right into them. I am very emotional - the tears started to flow. So did my mother's.

We went to the beach house. The group decided to go up to the ranch for dinner. Mose, who was in charge at the ranch, would prepare everything. One of the wives in attendance had met my mother in New York and called her on the telephone. She asked if it would be all right if I went to the ranch. I would be well chaperoned. We would probably stay over night.

I had a publicity picture shoot scheduled for the next day. Mother got Bill on the telephone. He said he would drive me to the location on the other side of the mountain in good time. My mother agreed to bring a change of clothes to the location.

We left the ranch early in the morning. Bill told his guests to go back to the beach house where he would meet them there after he dropped me off. He did not walk into the beach house until after six o'clock that night. He did not know where he had been. He had been driving all around. They took one look at him and knew this was it.

Three nights later he proposed. We were married three weeks to the day after we met, almost immediately after his divorce was finalized. The date, June 5, also was his birthday. He drove back from Lone Pine for the wedding that took place at my town house. We went to the beach house after the marriage. Mose had cake and champagne. We spent the night at the ranch and drove back to Lone Pine the next day so that he could be at work on Monday.

HLR: And, from that point forward, you were never separated?

GBB: The only times we were separated were the two nights that I left Biscayne to come back to get the house ready for Christmas. It was a holiday each time.

TOPPER

HLR: Tell me about Topper. How did Bill acquire him?

GBB: He heard about a horse, and he went out to look at him. This was shortly after we were married. We both fell in love with him. There was no question about it. This was the horse he wanted. I named him Topper after the Thorne Smith *Topper* novel I was reading at the time. [HLR Note: "Topper," a movie starring Norman McLeod, Constance Bennett, Cary Grant, Roland Young, and Billie Burke, appeared in 1937.]

HLR: Where were you living at the time Bill acquired Topper?

GBB: We had a ranch, located half way between Malibu and Oxnard, up the coast highway. It was a miniature San Simeon. The living room was fifty-six feet long, every piece of wood held in place by pegs. The stables were down below, built as part of the ranch.

HLR: Who cared for Topper?

GBB: Mike

HLR: From the beginning?

GBB: Initially, Mike was part time. He was one of the crew, one of the wranglers. Eventually he became Topper's personal groom. He was responsible for getting Topper to Hoppy's personal appearances and carrying for Topper.

HLR: Did you keep Topper at the ranch?

GBB: Mike took care of Topper at his place. He had a place in the San Fernando Valley. Mike would drive him to the locations.

HLR: You also had a horse, didn't you?

GBB: I received my horse during the first Christmas that we spent at the ranch. Bill had an outfit, saddle, and trappings made to match the one that he wore and used. I named the horse, Mrs. T. But she was too big for me, about as big as Topper. We replaced Mrs. T with an Indian pony. I do not know where they found him. I named this horse Turnabout. I kept him until we sold the ranch.

HLR: When Bill toured as Hopalong Cassidy, you had a special trailer made to transport Topper.

GBB: Yes. The trailer was custom made. It was a beautiful thing, black and white like everything else. Even the interior was black and white. There never was trouble getting Topper to go into the trailer. The trailer had port holes so Topper could look out.

Topper was not your average horse. When Topper saw something, he remembered. One of the things that made him such a remarkable animal was his curiosity and the fact that nothing bothered him.

HLR: When you took Topper on the road, where did he stay?

GBB: He stayed at local farms. Mike would stay with him. He made all the arrangements.

HLR: Was there anything special that Mike did to groom Topper?

GBB: Mike carefully filed Topper's teeth. He did not want anyone to know how old Topper was. Mike made certain that Topper always had a heavy undercoat and that his hooves were polished and shiny. He was beautiful.

Of course, there was a problem with sunburn. Topper had pink skin. He was almost albino, except for his bright blue eyes. The one thing that Mike never did was apply mascara to Topper's eyelashes. However, he did use Arden's eight hour cold cream on Topper's nose. Some of the parades and location shoots lasted for three or more hours. Topper had to be protected from the sun.

HLR: Bill and you obviously felt Topper was special. Are there any stories that illustrate this?

GBB: Topper was a clown. He always knew where the camera was. Hoppy always said you had better be sure you had your butt in the saddle at the end of a scene. Topper would stop dead the minute he was past the camera.

One time by accident, Topper was walking out of a scene. He accidentally stepped into a tin pail on the side of a road. It made this terrible sound. Of course, he started to stumble, and the scene had to be reshot. Everybody was hysterical. They thought it was very funny. That is all Topper had to hear. They removed the pail from the scene and did the scene over. Topper started to walk out. After a short distance, he deliberately walked out of the way and over to where they placed the pail and stuck his foot in it again. He just wanted another laugh.

Most horses do not remember. Topper did. All he needed to do was see how things worked once. When Hoppy and Topper where with the circus (Cole Brothers Circus), the first night Hoppy and Topper rode into the ring. All of sudden the lights came on. Topper just reared up and let out a scream. He was fine. The next night he knew it was going to happen, and it did not bother him a bit. He was like a big ham. He waited for the lights. I am not going to move until you put those spot lights on. He used his head. Topper thought.

HLR: You mentioned that Topper died right after a Rose Bowl parade. What year was that?

GBB: I think it was 1961. Bill frequently appeared in the Rose Bowl parade. It was tradition. He rode by himself, just Hoppy and Topper. Mike died about three weeks after Topper died.

HLR: What role did Topper's and Mike's death play in Bill's

decision to retire the Hopalong Cassidy character?

GBB: Bill felt that fate was sending him a message. In the first place, he could never ride another horse. He could never trust another horse the way he did Topper. Topper could be around thousands of children and not do anything.

Second, Bill wanted to walk away while he was at the top. He did not want to see Hoppy dwindle away. The Hoppy experience was so big, so wonderful, so beautiful. There isn't one word to describe it. It was remarkable. Bill wanted to leave it that way.

HLR: I think wisely so.

GBB: I do too.

On Location

HLR: What can you tell me about Hoppy's distinct costume?

GBB: The dark blue navy shirt was made like a leotard. It was a smooth body suit. The shirt fit under the pants so there was one smooth line. The pants were black pin stripes, very formal. They had stretch in them so Bill was able to move. But, they were skin tight.

The crease in the hat was an accident. While Bill was doing his first picture, he was standing there thinking about what he was going to do in the scene. He had the hat in his hand. Without realizing it, he was making a dent in the hat. They called him to do the scene. He did several scenes before realizing his hat had a big crease in the center. They decided to keep it. It became one of his trademarks.

HLR: Many of the Hoppy movies were filmed in Lone Pine. Any special remembrances?

GBB: Lone Pine was a good four hour trip, sometimes longer. Initially, we stayed in the Dow Hotel. There were no showers, only tubs. A lot of people where unhappy. There also was no food in the hotel. There was a couple of places across the street where you could get dinner. We normally sat with whomever was there.

Later we rented a little house in the Alabama Hills. Some prospector or someone had it. It was heated by kerosene. There was two inches of grease on the walls. We even found a hay rack in the orchard. Everything was so grown over. But, it was a beautiful spot. There was a natural pool. It was the only pool I have ever known where you could dive in and come out dry. The water was so cold.

Filming in Lone Pine was pretty dusty work. Temperatures often rose above one hundred degrees. The reflectors they used compounded that tremendous heat. The directors sat under umbrellas and had wet towels around their necks. The actors did not have any of that.

HLR: Were there ever any accidents on the set involving Bill?

GBB: One time the copper jacket of a bullet ricocheted off a rock and cut Bill right across his eye. They did not know whether he was blind or not. The only thing that saved him was the angle at which he was shooting. At the split second the jacket exploded, he had turned his eyes. The jacket cut him across the film of the eye rather than the lens of the eye. After the accident he said, "That's twice now. I'm not going to go for a third time."

One time while discounting from his horse, he stepped

on a small pebble. He turned and flipped over. His ankle just snapped. He returned to shooting as quickly as he could. They changed his outfit because he could not wear the tight pants and boots. He played a dude from the East and a lumberjack.

HLR: Gene Autry played an important role in your lives for a brief period. Tell me about it?

GBB: Gene Autry made his ranch available to us to film some of the Hoppy television half hours. It was a good location. It had Western streets with building fronts. Other filming was done in a little Hollywood studio.

There was one complete building at the Autry ranch, a little farm house with dining room, kitchen, and bedrooms. Autry graciously offered it to us, rent free. We stayed. Bill had been driving back and forth to the ranch to shoot. When he came home at night, it was dangerous. He reached levels of tiredness when he really should not have been driving.

Every two weeks we would get a weekend off. We went back to our house and stayed a couple of days. We followed that pattern until the series was shot. The farm house was a life saver. It was during the filming of this series that Bill had sun stroke.

Merchandising Copy

HLR: When Bill and you decided to go after the television and merchandising rights, did you have any idea of the potential?

GBB: Oh, no. No way. No way! All we were concerned with was what he would be doing with his life and the contribution that he could make to the Hopalong Cassidy char-

acter.

HLR: The results must have been overwhelming?

GBB: At that point of time, I do not think there was anything comparable. It was just phenomenal. Just phenomenal!

It was not just a kid thing, like Saturday morning television. You had parents who knew him from the films and grandparents who knew him as William Boyd before Hopalong Cassidy. You had mothers who thought he was the greatest. Men liked him. There was no feeling of envy or jealousy because their kids were crazy about Hoppy.

HLR: I have the impression that although there were thousands of products with Hoppy's name or logo, Bill personally felt strongly about the quality of the products with which he associated.

GBB: There were many fights about that. He turned down many things. He was adamant. He would rather they sell a tenth of what they could sell so long as that tenth was quality. I still have people come up to me and say, "I still have that Hoppy shirt that I had when I was a child."

Bill was really against the bubble gum when they first came to him. He said no. He did not want to be part of it. He did not think it was for the children, a terrible habit. I do not know how they persuaded him to do it.

Hoppy Cares

GBB: Whatever city we were in, Hoppy always visited the children's hospital. He was amazing. He always would go into the communicable disease wards. Bill knew the children wanted to see him in his Hoppy outfit. He refused to wear the customary jackets, hats, and masks.

Sometime I did not or could not go with him. What I saw often tore me apart. I was not chicken. I reached a point where I could not take it any more. He never did. He did them all. No matter where we were.

HLR: When did Bill's strong commitment to children's hospitals begin?

GBB: It was after we were married. He was making decisions about what he wanted to do with his life. He felt he owed his fame and fortune to children. He wanted to pay them back.

HLR: Is there any one story about Bill's hospital visits that you think symbolizes his commitment?

GBB: One time when we were in the Hollywood, Dorothy Lamour's mother called. Dorothy's son was very ill, unconscious. They were not certain he would live. The boy loved Hoppy. Dorothy's mother asked if Bill would visit the boy in the hospital. Bill came home, changed into the Hoppy outfit, and went to the hospital. The child had been in a coma for some time. Bill walked in and spoke to him. The child opened his eyes and said, "Hoppy." To this day, Dorothy Lamour swears that Bill saved the boy's life.

HLR: You mentioned that Bill was very concerned about children having to pay to see him?

GBB: His image was let the children come unto me. No one will have to pay. This led to a great many interesting stories involving his appearances at department stores and other promotional events. He made certain that anybody who wanted to see him, could see him.

This was one of the reasons why he was not happy during the season we spent with the Cole Brothers Circus. The circus seemed an ideal way to combine Hoppy, circus performers, animals, and kids. But, Bill was really unhappy because there was no way he could let everybody in free.

HLR: During our conversations you mentioned Bill's willingness to spend long hours standing in line to greet his many fans. Any particular memories?

GBB: Bill was one of the first to break the color barrier in the South. We were in Alabama on a department store tour. They had two lines - one for black children and one for white children. Bill said, "No way. No way. They come in together." You can image what happened when he said this. There was a lot of talk going on. This had never been done before, and they were not about to do it. But, they did. It worked. There were no riots. It was wonderful.

A Final Note

GBB: Bill used to think of Hoppy as his good side. Bill was a Gemini. Hoppy was not a fictional character that Bill played. Hoppy was an integral part of himself. He put himself into Hoppy. That's why I think Hoppy is so associated with him now.

You do not think of William Boyd as an actor who played Hopalong Cassidy. You think of them as one.

In the end, everyone just called Bill HOPPY.

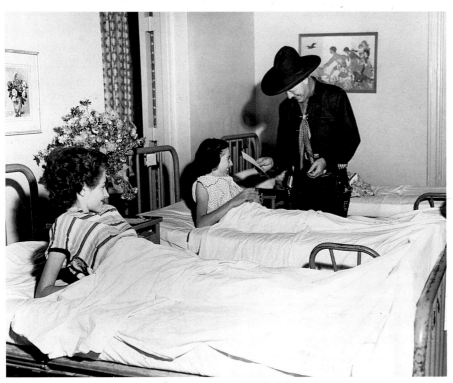

Chapter 4

HOPALONG CASSIDY COLLECTIBLES: STATE OF THE MARKET REPORT

The mid-1990s is a critical time for Hopalong Cassidy collectibles. The generation that revered Hopalong Cassidy as their favorite television cowboy hero is in their late forties and fifties. The individuals who idolized Hoppy as one of the great stars of the 1930s and 1940s B movie cowboy genre are retired and rapidly declining in numbers. The avid readers of Westerns novels who devoured the Hopalong Cassidy novels of Clarence Mulford are dead. For the thirtysomething and younger generations, the collectors of the 21st Century, Hoppy is a non-entity. Given these statistics, the news surrounding Hopalong Cassidy collectibles in the mid-1990s is surprisingly good.

I was fortunate to have rediscovered Hoppy in the early 1970s. I assembled the bulk of my collection between 1972 and 1982 when Hoppy material was plentiful and relatively inexpensive. I have watched Hoppy collectibles recycle in value several times in the intervening years. Currently, Hoppy is hot - at the peak of yet another collecting craze. Once again, he has proven that he is King of the Cowboy Merchandisers. The critical question is whether or not Hoppy is riding off toward greater adventures or into the sunset.

After a very strong collecting market in the 1970s, Hoppy collectibles experienced a lull in collector interest

during the 1980s. Roy Rogers and the Lone Ranger replaced Hoppy as the No. 1 cowboys among collectors. Roy was king-of-the-hill. Poor Gene. When it comes to cowboy collectibles, Gene Autry has always finished a distant fourth.

There were several reasons for the 1980s shift from Hoppy to Roy and the Lone Ranger. First, Hopalong Cassidy collectibles reached a pricing level by the end of the 1970s that made them appear extremely expensive when compared to memorabilia from other television cowboy heroes. Second, supply exceeded demand. The number of Hoppy collectors was relatively small, several hundred at most. The market had not matured to the point where generalist television cowboy collectors, crossover collectors, and decorating trends became major pricing influences. Third, Roy Rogers, Gene Autry, and to a certain extent the Lone Ranger benefitted from renewed exposure on television cable channels, video stores, and elsewhere. Hoppy was conspicuous by his absence. Roy and Gene still play a continuing role in the business and television industry. Each has been actively involved in the creation of a museum bearing their name. It is hard for a dead person to compete with living legends.

Just as the rise in value for Hoppy collectibles worked against Hoppy in the 1980s, the rise in value of collectibles associated with Roy, the Lone Ranger, and Gene worked in Hoppy's favor. By the late 1980s, their collectibles were at the same price level - no longer cheap, but still somewhat affordable.

A level price playing field is only one reason that led to a revival of interest in Hopalong Cassidy collectibles in the 1990s. The number of toy collectors increased exponentially in the 1980s. The toy market exploded. In the mid-1990s dozens of specialized toy periodicals, price guides, and shows exist to serve this market. The 1950s television cowboy heroes and their collectibles are part of this burgeoning toy market. This alone is enough to trigger a major revitalization of interest in Hoppy. The good news, at least in some collectors' and dealers' eyes, is that other major forces are at work as well.

Major collector focus on the post-World War II era increases with each passing year. There already are three generations of collectors who grew up after World War II. Almost one half of America's population cannot answer the question: "Where were you when you heard JFK was shot?" Hard though it is to accept, the late 1940s and 1950s are ancient history to the thirtysomething and younger collectors. Hoppy, Roy, Gene, and the Lone Ranger are hovering on the brink of becoming antiques.

The key is television. Television is the instrument that currently separates the antiques from the collectibles collector. Since Hopalong Cassidy was one of the pioneers of early television, he is attracting renewed attention from collectors who want to focus on this period as a whole, not necessarily on one character. You cannot have a collection of early television memorabilia without several Hopalong Cassidy items.

The generalist cowboy character collector, i.e., a person who collects memorabilia from the television cowboy shows of the late forties through the early sixties, is a recent arrival on the collecting scene. His focus is on the genre. The guiding principle is the earlier, the better. Once again, who benefits? Hoppy.

Perhaps the biggest market influence is the crossover collector. Collectors of advertising characters, Aladdin lamps, bicycles, cap guns, cereal boxes, cereal premiums, character watches, character radios, Silver Age comic books, milk bottles, movie posters, non-sport trading (bubble gum) cards, and pocket knives have driven the price of Hoppy licensed products in these areas through the roof. This list is by no means inclusive. Almost every Hoppy items crosses over into one to six additional collecting categories. There may be more Hopalong Cassidy material in outside collections than in Hopalong Cassidy collections.

The 1990s are witnessing a revival in the Western look. Decorators favor decors that feature artifacts ranging from items used by real live cowboys to pieces with Western motifs and images of the 1950s dude ranch Western. Maple bedroom suites are hot. Using products associated with early television cowboy heroes, especially if they are pizzazz oriented, is encouraged. The highly colorful Hoppy tin lithographed toys and textiles licensed by William Boyd fit the bill.

One of the great unanswered questions among collectors and dealers is: how much Hopalong Cassidy merchandise has survived? While most greatly underestimate the amount, the truth is that fewer and fewer Hoppy items are showing up at garage sales and house auctions. There still is a great deal of material awaiting discovery. However, the peak has passed. Each year brings less and less new material into the market

Finally, Hoppy's general public exposure is increasing thanks to the efforts of U.S. Television Office and others. A video release program featuring complete and remastered Hopalong Cassidy movies and televisions shows is in its early stages. The annual Hoppy Festival in Cambridge, Ohio, museum exhibits, such as the one at the American Museum of the Moving Image in 1994-95, and books, such as this one, are introducing Hoppy to a new generation and rekindling strong nostalgic urges in older ones. U.S. Television Office has scored several recent successes in licensing new Hoppy products and continues to pursue the making of a new Hopalong Cassidy motion picture.

Little wonder Hopalong Cassidy collectibles are enjoying a resurgence of interest. The support is multi-layered. Recently, Ted Hake, owner of Hake's Americana and Collectibles, America's leading mail auction and author of *Hake's Guide To Cowboy Character Collectibles* (Wallace-Homestead, 1994), confirmed that Hoppy is one again King of the Cowboy Mechandisers. Prices across the board are at record levels. Common pieces have doubled or tripled in value over the past five years. Scarcer pieces have risen ten times or more.

The news could not be better for a Hopalong Cassidy devotee. Their hero is king-of-the-hill. What more could anyone ask? Shoot the nay sayers.

I confess that I am one of the nay sayers. Amidst this overwhelming amount of good news, I see disaster. Allow me to state my case before you decide to hang me.

First, the price of many Hoppy items has risen to the point where they are no longer affordable to the average collector, even if Hoppy was his childhood hero. Recently someone offered me a Hopalong Cassidy night light (the Aladdin gun and holster model) for $400. It books in *Hake's Guide To TV Collectibles* (Wallace-Homestead, 1990) for $150. Did it more than double in value since 1990? Absolutely, not. Further, whether it is $150 or $400, this a great deal of money. How big a collection can one build when the average unit price is in the fifty to one hundred dollar range? There are a large number of rustlers among the sellers of Hopalong Cassidy collectibles and no sheriff in sight to round them up and take them to the hoosegow. When a collectible category is no longer perceived as affordable, its values decline and then stabilize. The Hoppy pricing structure already has reached this point.

Second, are Hopalong Cassidy collectibles a one generation collectible? What interest will exist in Hopalong Cassidy collectibles when the generation that grew up with Hoppy (a) stops collecting, (b) sells, and (c) dies. Most collectors cease collecting in their early sixties. Many Hoppy collectors are in their sixties. Specialty collectors sell in their late sixties or early seventies. As long as there is a younger generation of collectors waiting to buy, the market sustains itself.

What happens when there is no younger generation of collectors? After a few strong sales (those who abandon ship first are saved), the market collapses. Supply exceeds demand. I believe that there is an excellent chance that high-ticket Hoppy items purchased today will sell for nickels and dimes on the dollar in 2020, twenty-five years from now. Hopefully, I will live long enough to take advantage of the situation.

My view is a minority one and not shared by most Hoppy collectors and dealers. My advice is to them is to look at themselves in the mirror and to carefully note the color of the hair of their customers. The simple truth is that the vast majority of buyers of Hoppy material are over fifty. This is not a positive sign.

What will it take to prove me wrong? The answer is a major renewal of interest in the television Western and, ideally, a Hoppy revival as part of it. The television Western lives, but on the program edge, not in the center. Each new television Western series, television special, or movie raises my hopes. Each failure depresses me a little more.

The news in 1994-95 was not good. Maverick and Wyatt Earp were box office failures. The Cisco Kid movie on TNT was well received, but did zilch for Cisco Kid collectibles. "The Adventures of Brisco County, Jr." did not survive its first full season. The 1990s remain the era of family sitcoms and space adventure shows. Will Lonesome Dove, the television show, turn the tide? I doubt it.

Perhaps, it is time for Western devotees to face the truth. The frontier of the 21st Century is outer space, not Planet Earth. We are a vanishing breed, like the buffalo. We will survive, but in small herds in fenced-in preserves. In a world of mass production, we will be overwhelmed trying to save all the objects that have survived. We will lend credence to the old adage: "Just because it is old does not mean that it is worth money."

Chapter 5

SELLING HOPPY

By early 1950s over a hundred manufacturers were producing Hopalong Cassidy licensed products. At its peak, the Hoppy craze produced annual retail sales between 50 and 70 million dollars. William Boyd became a millionaire several times over. Americans of all ages could not get enough of Hoppy.

Few American households escaped the Hoppy phenomenon. Hoppy products touched every aspect of a child's life - from the time he arose in the morning until he went to sleep at night. Hoppy-licensed merchandise sold tens of thousands, hundreds of thousands, and even millions of units in an era when thousands of units was considered acceptable. While not every Hoppy license brought success to its manufacturer, over ninety percent did.

Boyd's focus on the food industry, a lesson learned from Disney, insured his popularity with parents as well as children. Although William Boyd licensed Hoppy to a wide range of food products, Hoppy is most identified with the bakery and dairy industries. Milk and bread were two of the major food groups considered essential for a growing child in the 1950s. Producer's Dairy of Fresno, California, still uses Hoppy to promote its milk products. This relationship is one of the longest continual character licensing agreements in history - perhaps the longest.

Most Hoppy collectors focus on licensed products from the 1949 to 1959 period. Very much overlooked are the Clarence Mulford novels and pre-Hoppy William Boyd material. While Hoppy collectors may own one or two examples of a Hopalong Cassidy movie poster or lobby card, most shy away from these objects because of their high cost, a result of the high prices paid by movie poster collectors and members of the decorating community.

What follows is a sampling of Hopalong Cassidy collectibles. It will introduce you to the various collecting categories within the wonderful world of Hoppy collectibles and allow you to cast your cares aside while leisurely strolling down nostalgia lane. If you are old enough to have grown up with Hoppy, you will think "I owned one of those" or "I remember one of those." If you are not, it is my fondest wish that your reaction from time to time will be - "Wow! That's neat. I won't mind owning one."

A comprehensive price guide to Hopalong Cassidy collectibles is found in Appendix 6. This approach was necessary to (1) save space and (2) avoid a myriad of similar images. While the Hoppy devotee would delight in seeing every movie poster, movie lobby card, bubble gum card, comic book, and milk bottle pictured, such an approach would put the average user of this book to sleep. Use the price guide to find the value of your Hoppy collectibles. Use the pictorial section to understand the breadth of Hoppy's influence.

Since the pictures represent broad collecting categories, the captions focus on collecting hints and information. Used in conjunction with the price guide, this allows insights into what drives the Hoppy collectibles market and why — something not generally found in books of this type. If all you do is look at the pretty pictures, you will miss a good portion of the value of this book. As Hoppy would say, "Pay Attention."

Silver Screen Hoppy

Although a few Hopalong Cassidy movie three sheet posters, measuring approximately 41" x 81," are known, the vast majority of Hopalong Cassidy movie posters currently available on the market are one sheets, measuring approximately 27" x 41". In addition to one sheets for the initial release of a Hoppy movie, there are one sheets for re-releases (look for an **R** after the year of re-release) and the re-released and renamed North Vine series. The renamed movies are as follows:

Original Title	Renamed Title
Devil's Playground	*Gunpowder Valley*
Fool's Gold	*The Man From Butte*
Unexpected Guest	*Saddle and Spurs*
Dangerous Venture	*Six Shooter Justice*
Hoppy's Holiday	*The Fighting Cowboy*
The Marauders	*King Of The Range*
Silent Conflict	*Trigger Talk*
The Dead Don't Dream	*Fighting Man From Arizona*
Sinister Journey	*Two Gun Territory*
False Paradise	*The Fighting Texan*
Strange Gamble	*Stampede Fury*
Borrowed Trouble	*Law Of The Trail*

A large number of Hopalong Cassidy inserts, approximately 14" x 36", have survived. These tall, narrow posters, usually printed on a heavier stock than the one sheets, often are far more graphic than the design of the one sheet. They normally retail between 25% and 50% less than the value of the one sheets for that same title.

Most Hoppy lobby card sets contained eight cards, measuring approximately 11" x 14". An eight card set usually consisted of seven scenic cards and one title card. Paramount, the distributor of most of the Hoppy films, often did not use a title card, replacing it with a close-up or portrait card instead.

Press books have been located by collectors for most Hopalong Cassidy movies, both initial and re-releases. Hoppy collectors favor the re-release press books because they often included promotional material for Hopalong Cassidy licensed merchandise. Many press books contain loose sheets that often become separated from the press book. Do not pay a premium price for an incomplete unit.

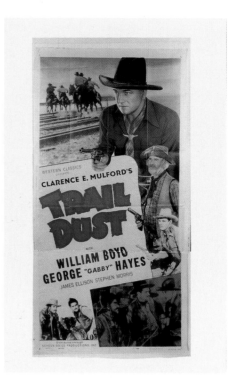

Poster, three sheet, *Trail Dust*, released December 19, 1936. George "Gabby" Hayes appeared in twenty-two Hopalong Cassidy films, playing the character Windy Halliday in nineteen of them. *Trail Dust* was his eighth Hoppy film.

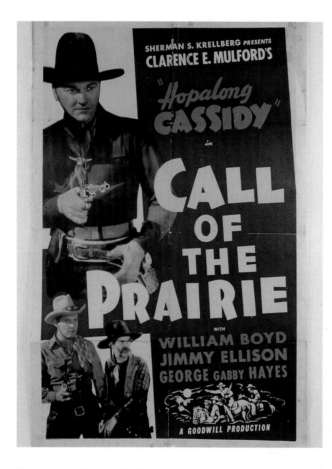

Poster, one sheet, *Call of the Prairie*, released March 6, 1936. Jimmy Ellison played the role of Johnny Nelson in eight of the first nine Hopalong Cassidy movies.

Poster, one sheet, *Silver On The Sage*, released March 31, 1939. Russell Hayden first appeared as Lucky Nelson in *North of the Rio Grande*, released on June 18, 1937.

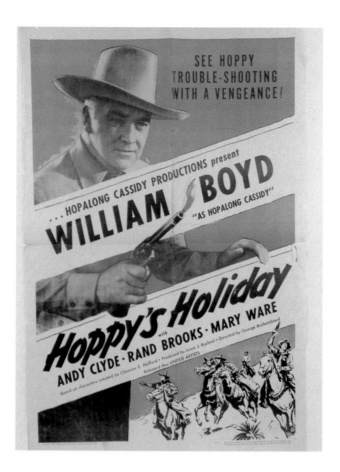

Poster, one sheet, *Hoppy's Holiday*, released July 18, 1947. The last twelve feature length motion pictures are know as the North Vine series and produced by William Boyd Productions.

Insert, *Silent Conflict*, copyrighted March 19, 1948. The North Vine Hoppy films followed the trio western format, featuring William Boyd as Hoppy, Andy Clyde as California Carlson, and Rand Brooks as Lucky Jenkins.

Inserts for *Border Vigilantes*, released April 18, 1941, *Lumberjack*, released April 28, 1944, and *Sinister Journey*, released June 11, 1948. *Border Vigilantes* was Andy Clyde's fourth Hopalong Cassidy film appearance. Clyde brought a level of consistency to the comic side kick role in Hopalong Cassidy movies that had been missing since the departure of George "Gabby" Hayes.

39

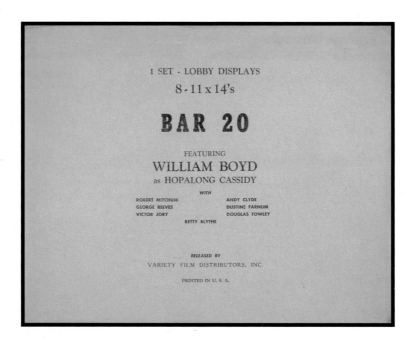

1 SET - LOBBY DISPLAYS
8 - 11 x 14's

BAR 20

FEATURING
WILLIAM BOYD
as HOPALONG CASSIDY

WITH

ROBERT MITCHUM   ANDY CLYDE
GEORGE REEVES   DUSTINE FARNUM
VICTOR JORY   DOUGLAS FOWLEY

BETTY BLYTHE

RELEASED BY
VARIETY FILM DISTRIBUTORS, INC.

PRINTED IN U. S. A.

Lobby Card set envelope, *Bar 20*, released October 1, 1943.
Lobby card set envelopes are not common. They add an
additional ten to twenty percent to the value of a lobby card
set.

Lobby Cards for *Sinister Journey*, released June 11, 1948, and
Dangerous Venture, released May 23, 1947. William Boyd
Productions completed shooting most of the North Vine
features in six days, a typical schedule for a "B" movie western
of the times.

Lobby Cards for *Unexpected Guest*, released March 28, 1947, and
Forty Thieves, released June 23, 1944. The general consensus
among film critics is that the North Vine Hopalong Cassidy
movies produced by William Boyd Products do not match the
quality of those produced by Harry Sherman Productions.

Lobby Card, *Jornada Sangrienta (Sinister Journey)*, released June 11, 1948. The English title above the photograph reads "Two Gun Territory," the re-release name for this film. The lobby card proves that like many modern films, Hopalong Cassidy films enjoyed a second life in foreign release

Poster, One Sheet, *Oro De Mala Ley (Fool's Gold)*, released January 31, 1947. While comic and romantic sidekicks along with an assortment of villains often repeated in Hopalong Cassidy films, female leads did not.

Lobby Card, *Justicia A Trio Limpio (Dangerous Venture, a.k.a., Six Shooter Justice)*, released May 23, 1947. Hopalong Cassidy films enjoyed a worldwide audience. In addition to Mexico and elsewhere in Central and South America, Hoppy films were extremely popular in England, France, the Scandinavian countries, South Africa, and Japan.

Lobby Card, *A Galope Tendido (Unexpected Guests, a.k.a., Saddle and Spur)*, released March 28, 1947. Patricia Tate also appeared in Hoppy's next film, *Dangerous Venture*.

Lobby Card, *"El Diablo Anda Suelto, (Devil's Playground, a.k.a., Gunpowder Valley)*, released November 15, 1946. *Devil's Playground* was the first film of the North Vine series.

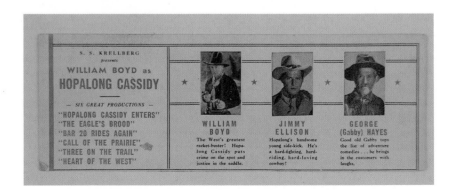

Ink Blotter, S. S. Krellberg. This blotter touts the first six Hopalong Cassidy films that William Boyd made. Boyd received $30,000 for his six performances and a new lease on life.

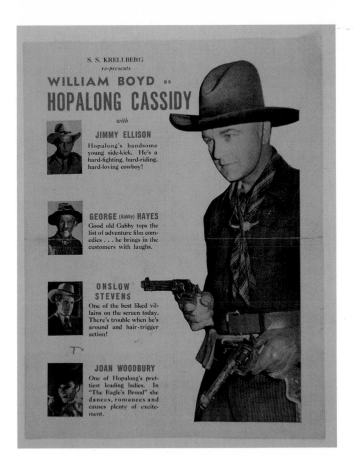

Promotional Folder, S. S. Krellberg, four fold. Joan Woodbury (Nana Martinez) starred in only one Hopalong Cassidy film. Onslow Stevens found fame in other film vehicles.

Press Book, *Texas Trail*, released November 26, 1937. Sherman attempted several different approaches to the Hopalong Cassidy character in the 1935 and 1936 films. By 1937 the character evolution was complete. Hoppy's unique costume coupled with a personality that balanced personal strength and wisdom with tenderness developed a large following among "B" movie cowboy aficionados.

Press Book, *Trail Dust*, released December 19, 1936. Early press books focused primarily on advertising and theater promotions. Sherman apparently made little or no effort to license Hopalong Cassidy merchandise, choosing not to following efforts by Gene Autry and other "B" movie cowboys.

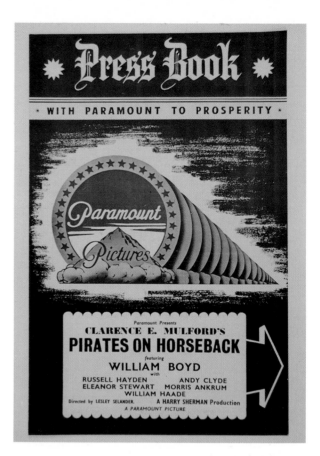

Press Book, *Pirates on Horseback*, released May 31, 1941. By 1941 Sherman had exhausted the Mulford novels as a source of inspiration for Hopalong Cassidy story lines. Ethel La Blanche and J. Benton Cheney wrote the screenplay for *Pirates on Horseback*. Cheney worked on a total of nine screenplays for Hoppy movies.

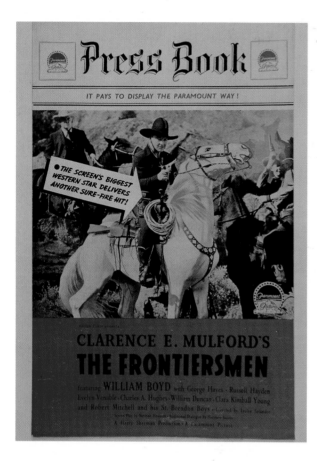

Press Book, *The Frontiersman*, released December 16, 1938. Paramount distributed the Hopalong Cassidy films released between 1935 and 1941. In 1942 United Artists assumed distribution.

Press Book, United Artists. In addition to individual press books, United Artists issued a general "Hopalong Cassidy Publicity and Exploitation Manual." As with the individual press books, the emphasis was on theater promotions, e.g., contests, rather than merchandise.

Promotional Booklet, Capitol Film & Radio Co., four pages.
Sherman released *Forty Thieves,* his last Hopalong Cassidy
film, on June 23, 1944. William Boyd Productions released *The
Devil's Playground*, the first of the North Vine films, on November 15, 1946. In order to keep Hoppy before the general public,
Sherman signed an agreement with Capitol Film & Radio
Company to release thirty-five Hoppy films for home and club
showing.

Promotional Booklet, Capitol Film & Radio Co., four pages.
Hayes and Boyd may have finished "2nd and 3rd place at the
Box Office!" in a poll, but, not in 1946. Capitol had no qualms
about using 1930s statistics in its mid-1940s promotion.

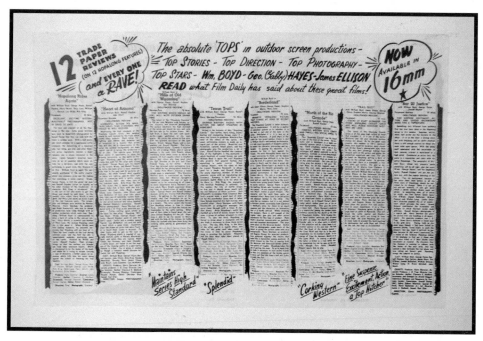

Promotional Booklet, Capitol Film & Radio Co., four pages.
The 16mm format allowed the films to be shown in venues
other than movie theaters. Hoppy became a popular feature at
neighborhood block parties.

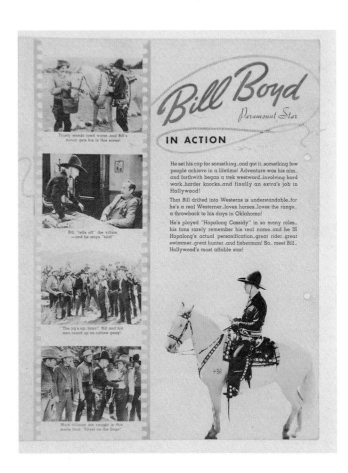

Original Title	New Title
Bar-20	Hopalong Cassidy's Rustler Round-Up
Bar-20 Days	Hopalong Cassidy's Private War
The Coming of Cassidy - and the Others	The Coming of Hopalong Cassidy
Hopalong Cassidy's Protégé	Hopalong Cassidy's Saddle Mate
The Bar-20 Rides Again	Hopalong Cassidy's Bar 20 Rides Again
Trail Dust	Hopalong Cassidy With The Trail Herd

Notebook Filler Page. Boyd's successful portrayal of Hopalong Cassidy made him a recognizable Hollywood personality. It is common to find Boyd as one of the featured artists in any 1940s or 50s series material with a Hollywood star theme.

Read All About Him

Clarence Mulford's first Hopalong Cassidy story appeared in the December 1905 issue of *Outing* magazine. Several additional Hoppy stories followed. In 1907 *Outing* published Mulford's short story Hoppy series in novel form as the *Bar-20*. In 1909 *Outing* was sold, and its book publishing discontinued.

Mulford moved to A(lexander) C(aldwell) McClurg, who published a series of Hoppy novels beginning with *Hopalong Cassidy* in 1910 and ending with *Tex* in 1922. In 1924 Mulford left McClurg and moved to Doubleday. *Hopalong Cassidy Returns*, published in 1924, was Doubleday's first Hoppy novel. Mulford's final Hopalong Cassidy book for Doubleday was *Hopalong Cassidy Serves A Writ*, published in 1941.

Many of Mulford's Hopalong Cassidy novels were issued in inexpensive, hard cover re-issues. A. L. Burt used the same plates as the original. In the late 1930s Mulford switched his allegiance to Burt's principal competitor, Grosset & Dunlap. In addition to these low-priced editions, Doubleday's reprint house, Garden City, also re-printed several of the Mulford Hoppy novels.

In 1950 Grosset & Dunlap published several of the classic Hopalong Cassidy stories with new titles. The list is as follows:

Dell, Graphic, Harlequin, Pocket, Popular, and Western Novel Classics, a digest publisher, issued Mulford Hoppy novels in paperback, e.g., Graphic #23, *The Man From The Bar 20*, published in 1950. Three titles, *The Man From The Bar 20*, *The Bar-20*, and *Tex*, were published in special Armed Services Editions during World War II and sent to soldiers serving overseas.

Two areas ignored by American Hopalong Cassidy collectors are Clarence Mulford's other western writings, both short stories and novels, and English and foreign language editions of Mulford's Hoppy novels. Foreign language editions were published in France, Germany, Czechoslovakia, Denmark, Norway, Mexico, Italy, Finland, and Poland. The National Library for the Blind, Manchester, England, issued *Bar-20 Days* and *Bar-20 Three* in Braille editions.

In a desire to capitalize on the Hoppy craze of the 1950s, the Mulford estate contracted with Louis L'Amour to write four pseudonymous Hopalong Cassidy novels. Doubleday published four new Hoppy novels - *Rustlers Of West Fork*, *Trail To Seven Pines*, *The Riders Of High Rock*, and *Hopalong Cassidy, Trouble Shooter* - by Tex Burns in 1951 and 1952. Hodder & Stoughton published the English editions.

Bonnie Books, Cozy Corner Books, Doubleday, Samuel Lowe, and Western Printing and Lithograph Company (Little Golden Books) produced a variety of Hopalong Cassidy titles directed toward juveniles in 1950 and 1951. Several books had novelty features such as pop-ups (known as Hop-ups) and television wheels.

In 1943 Fawcett Publications began a *Hopalong Cassidy* comic book series that was to last 135 issues. Fawcett published No. 1 (1943) through 85 (1954). National published issues 86 (1954) through 135 (1959). Illustrators included Ralph Carlson, Gene Colan, Gil Kane, and Joe Kubert. Hoppy comics also appeared as part of Fawcett's *Master Comics*, *Real Western Hero*, *Six-Gun Heroes*, and *Western Hero* series. Bill Boyd appeared as himself in *Bill Boyd Western*, a twenty-three comic book series published by Fawcett between 1950 and 1952, and in Fawcett's *Western Hero*, No. 87 (1949) through No. 95 (1952).

In 1950s the Los Angeles Mirror Syndicate began the national distribution of a Hopalong Cassidy newspaper strip. The artist was Dan Spiegle. Royal King Cole did most of the writing. In 1951 King Features acquired the strip from the Los Angeles Mirror Syndicate. At its peak, the strip was in 200 plus newspapers. Spiegle and Cole ended the strip in 1955.

Bernard A. Drew's *Hopalong Cassidy: The Clarence E. Mulford Story* (Metuchen, NJ: Scarecrow Press: 1991) and Francis M. Nevins, Jr.'s *Bar-20: The Life of Clarence E. Mulford, Creator of Hopalong Cassidy, with Seven Original Stories Reprinted* (Jefferson, NC: McFarland & Company: 1993) provide complete lists of all of Mulford's published material, Hopalong Cassidy related or otherwise. They are must own books for any Hoppy collector.

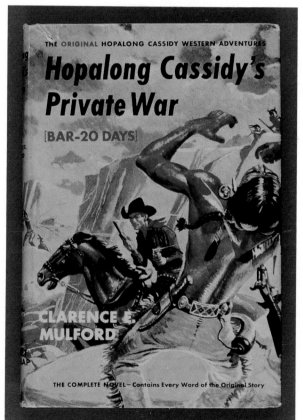

A selection of Clarence Mulford's Hopalong Cassidy novels from a variety of publishers. Hopalong Cassidy collectors are concerned primarily with owning a copy of each novel not whether or not it is a first edition, first printing, inexpensive edition, or retains its dust jacket.

A selection of Hopalong Cassidy juvenile, paperback, and pulp Hoppy publications. Collectors eagerly seek the juvenile titles, but not the paperbacks and pulps. The reason is simple. The juveniles feature illustrations of the Hoppy character that they know and love.

Novel, *Hopalong Cassidy's Private War*, a 1950 Grosset & Dunlap edition of *Bar-20 Days*. Expect for the Tex Burns novels and a Grosset & Dunlap 1950 edition of *Hopalong Cassidy*, none of the novel dust jackets or interior illustrations used Boyd's Hopalong Cassidy image for illustration purposes.

Novels, *Hopalong Cassidy en El Misterio Del Rancho Desaparecido* and *Hopalong Cassidy Y Los Cuatreros De West Fork (The Rustlers of West Fork)* by Tex Burns. These novels were published in Mexico by Editorial Constancia in 1952.

Juvenile, *Hopalong Cassidy and Lucky at the "Double X" Ranch*, Doubleday, 1950. William Boyd Enterprises, Inc., was not responsible for book licensing. This responsibility remained with Doubleday and/or the Mulford estate.

Juvenile, *A Television Book of Hopalong Cassidy and His Young Friend Danny*, illustrated by Jack Crowe, Bonnie Books, 1950. Samuel Lowe published this book without the television window as *Hopalong Cassidy and His Friend Danny*.

Juvenile, *Hopalong Cassidy and Lucky at Copper Gulch*, Garden City Publishing, 1950. Jack Crowe also illustrated this large size television book.

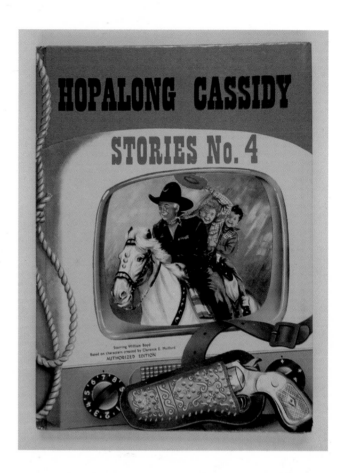

Juvenile, *Hopalong Cassidy Stories No. 4*, London, Adprint, ca. 1956-57. Contains six short stories written by Charles Hitchcock and illustrated by George Shaw and Charles Bourbe.

Juvenile, *Hopalong Cassidy Stories No. 3*, London, Adprint, ca. 1955-56. Contains eight short stories written by Charles Hitchcock and illustrated by G. A. Facey.

50

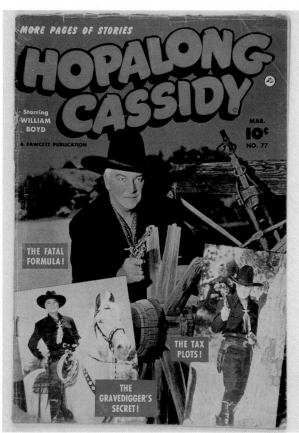

Juvenile, *William Boyd: Star of Hopalong Cassidy: On The Range*, No. 517-5, and *William Boyd and his friend Danny*, No. 518-5, Samuel Lowe. Lowe received a license from William Boyd Enterprises for coloring books and from Doubleday for story books. Both formats were used in these small size children's books.

Comic Book, *Hopalong Cassidy*, Fawcett, No. 77. Hopalong Cassidy merchandise frequently was advertised in his comics. The clothing advertisements are especially helpful in identifying which pieces of clothing were made by specific manufacturers.

Comic Book, *Hopalong Cassidy*, Fawcett, No. 49. After the first few issues that featured cartoon covers, most Hopalong Cassidy comic books had a picture of Hoppy along with Topper on the cover. In the comic books, Hoppy was the sheriff of Twin River County.

Comic Book, *Hopalong Cassidy*, Spanish edition. Hopalong Cassidy comic books enjoyed great popularity in Central and South America and in Europe. Only rarely do copies appear in the American market.

Premium, comic book, Hopalong Cassidy, Post Grape Nuts Flakes. Comic book premiums were used by Bond Bread, Post's Grape Nuts Flakes, and White Tower. Examples in their period packaging are scarce.

Magazine, *Life*, June 12, 1950. By mid-1950 the Hoppy craze had swept the nation. Boyd as Hoppy was featured on the cover of over a dozen national publications.

Magazine Tear Sheet, Life, June 12, 1950. Printed information about the Hopalong Cassidy merchandising craze is scarce. What little is known comes from magazine articles such this. Note the wide variety of clothing options available. The cheapest outfit cost $4.95, the most expensive $45.20.

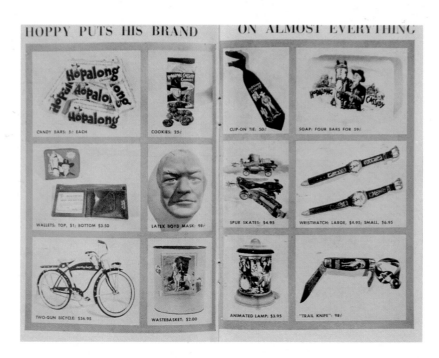

Magazine Tear Sheet, *Life*, June 12, 1950. Collectors constantly ask: "What did that cost originally?" Some of the prices are: 25¢ for box of Burry's Hopalong Cassidy Cookies, 59¢ for four bars of Hoppy soap, $2.00 for a wastebasket, $4.95 for a pair of Hoppy Spur Skates, and $56.95 for a Hopalong Cassidy bicycle.

Magazine, *Look*, August 29, 1950. Roy Rogers may have been "King of the Cowboys"; but, Hoppy was "Public Hero No. 1." Strong support from parents was one of the keys to Hoppy's success.

Magazine Tear Sheet, *Look*, August 29, 1950. The popularity of "B" movie westerns on television of the late 1940s and early 1950s made Americans Western conscious. Every child wanted to be a cowboy or cowgirl. One of the popular interior decorating styles of the period was dude-ranch western.

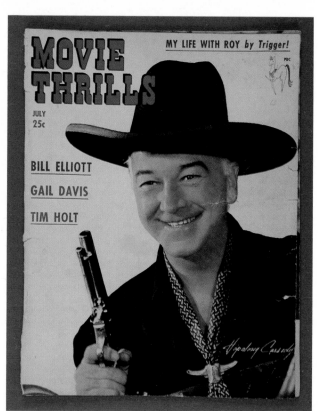

Magazine Tear Sheet, *Look*, August 29, 1950. While some children were intensely loyal to a single television cowboy star, most simply mixed and matched. *Look* priced out a typical first grader's cowboy wardrobe at $144.20, not cheap by 1950 standards.

Magazine Tear Sheet, *Look*, July 18, 1950. Dozens of magazines without a television cowboy image on the cover featured articles about Hopalong Cassidy and the general cowboy craze of the early 1950s. As reluctant as Hoppy collectors are to admit it, Gene, Roy, and others also helped fuel the fire.

Magazine, *Movie Thrills*, July 1950. The movie industry identified strongly with the successful transition of Hopalong Cassidy from movies to television. Theaters continued to offer full length versions of the movies for those who wanted to view the portions cut for television presentation.

Magazine, *Quick*, May 1, 1950. Quick was one of the first national magazines to report the Hopalong Cassidy craze. The others followed quickly thereafter.

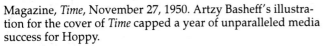

Magazine, *Time*, November 27, 1950. Artzy Basheff's illustration for the cover of *Time* capped a year of unparalleled media success for Hoppy.

Magazine, *TV Digest*, November 29, 1952. *TV Digest* was a forerunner of *TV Guide*. Editions were done on a statewide or regional basis.

Magazine, *Time*, November 27, 1950, Canadian edition. The nationalistic focus of most American collectors prevents them from appreciating the international aspects of the Hopalong Cassidy craze. Hoppy's popularity in Canadian was every bit as strong as it was in America.

Magazine Tear Sheet, *TV Digest*, early 1950s. Children not only loved Hoppy, but Topper, his horse, as well. Topper's gentleness around children added an extra dimension to Boyd's public appearances.

Magazine Tear Sheet, *TV Guide*, late 1950s. Hopalong Cassidy no longer appeared as a regularly featured show in New York, Philadelphia, Chicago, or Los Angeles by the time this article was written. However, stations in other cities that continued to air Hoppy films paid $1,000 to $1,250 per showing. As the article states: "That ain't hay."

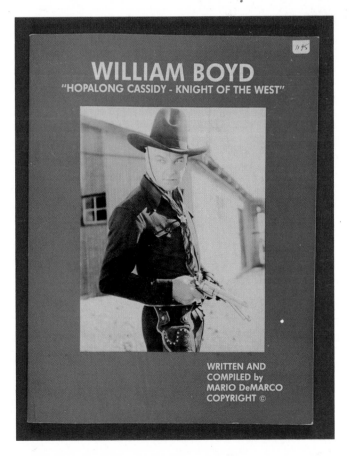

Promotional Photograph, *TV Guide*. No album series of 1950s cowboys would be complete without a picture of William Boyd in his role as Hopalong Cassidy.

Book, Mario DeMarco, *William Boyd: "Hopalong Cassidy - Knight of the West,"* published by the author. One of series of books DeMarco did about "B" cowboys. The format consists primarily of stills interspersed with text. The book's value rests in personal interviews with Pierce Lyden and Fred Scott.

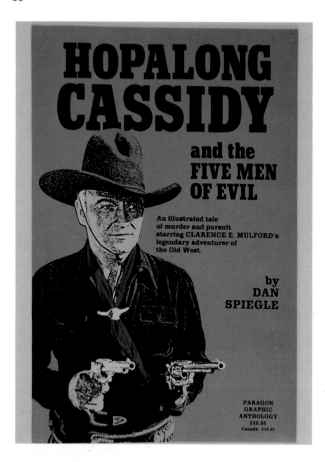

Dan Spiegle, *Hopalong Cassidy and the Five Men of Evil*, Paragon Graphic Anthology, a reprint of one of Spiegle's Hopalong Cassidy cartoon series. This is one of the many new products resulting from the aggressive licensing of Hopalong Cassidy material by U.S. Television Office, Inc.

Magazine, *favorite Westerns*, Issue #8, January 1983. In 1983 *favorite Westerns* was a serial magazine published six times a year by Serial World Publishers of Mankato, Minnesota. This issued features Bob Pontes' article "Hopalong Cassidy — The Man in Black."

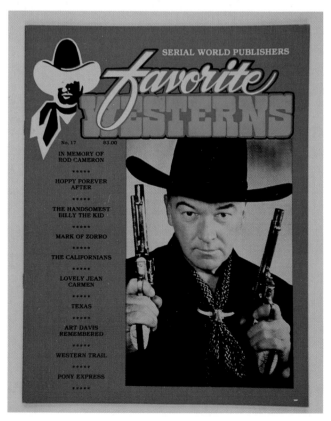

Magazine Tear Sheet, Hopalong Cassidy cartoon strip. Hopalong Cassidy enjoyed a strong following in the Scandinavian countries, particularly Denmark.

Magazine, *favorite Westerns*, Issue #17, 1984. In 1984 *favorite Westerns* became a quarterly. This issue contains Neil Summers' article "William Boyd - Hoppy Forever After."

Merchandizing Hoppy

By the early 1940s, William Boyd had become so closely associated with the character of Hopalong Cassidy that it was impossible to perceive him in any other role. The prime reason that Boyd founded William Boyd Productions and produced the Northvine Hoppy films was the lack of interest in him by studios for other lead roles. Boyd was in his late forties at the time.

Until the Hoppy craze of the early 1950s Boyd tried to separate his private life from his public image as Hopalong Cassidy. He lived and dressed stylishly. He participated actively in the Hollywood social scene.

Occasionally he appeared publicly to promote the Hopalong Cassidy films. However, no formal promotional tour was held. Remember, the Hopalong Cassidy movies were "B" movies. In the mid- to late 1940s Boyd made guest appearances on the rodeo circuit as a means of supplementing his income. Programs from these appearances have survived. When Boyd appeared in advertising campaigns, it was in civilian dress. Sherman owned the commercial character rights to Hopalong Cassidy until Boyd purchased them.

Everything changed once Hopalong Cassidy movies began to appear on television. Boyd recognized the need and made a strong personal commitment to support the television series and his licensed merchandise. Hoppy's image appeared everywhere, from advertisements for licensed products to endorsements for non-licensed products. Hoppy could sell anything.

Boyd set strict standards for the products he licensed and endorsed. He was committed to the principle that buyers would receive value for money spent. He turned down many offers because the goods did not meet his standards either in quality of production or final sale price. Boyd's licensing contracts called for final approval of all products prior to manufacturer. In reality, many manufacturers sought advise from Boyd during the development stage. This avoided last minute delays when production was ready to being.

Boyd relied heavily on Marguerite Cherry, Dan Grayson, Lew Pennish, and Robert Stabler to oversee the licensing activities of Hopalong Cassidy, Inc., Hopalong Cassidy Company, and eventually Hopalong Cassidy Enterprises' Merchandise Division. They handled the paper work, from contracts to ascertaining that royalties were paid promptly and accurately.

Contracts with numerous premium manufacturers are a little understood aspect of the Hopalong Cassidy merchandising program. Not all Hopalong Cassidy items were purchased in stores. Many bakeries, dairy companies, newspapers, and television stations used Hoppy premiums as giveaways during promotional contests. Hoppy licensed Banthrico Industries, Inc./Bar-Twenty Associates to supply the material needed by the hundreds of the banks that participated in the Hopalong Cassidy Bank Savings Plan.

The merchandising of Hoppy began in mid-1949 and effectively ended by the late 1950s. The Golden Age was 1950 through 1954. Product licenses such as clothing and play things dominated the first part of the Golden Age. Food products, e.g., milk, ice cream, etc., were strongest in the second part.

By the early 1960s the generation that grew up with Hoppy on television were young adults whose interests were directed elsewhere. Hoppy slowly faded from public view. It was not until the mid-1970s and early 1980s that they nostalgically renewed their contact with their childhood cowboy hero and began collecting Hopalong Cassidy merchandise in earnest.

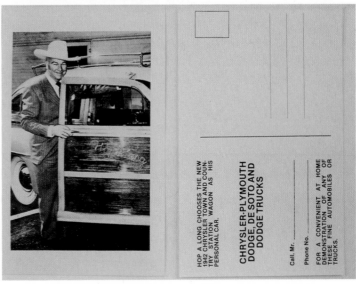

Post Card, advertising, Chrysler, 1942. Note Boyd's appearance - business suit and white hat. Boyd preferred being himself, rather than Hoppy when making public appearances and endorsements during the production of the Hopalong Cassidy films.

Stamp Set, "Hollywood: Official Stamps of the Stars and Studios," Set Q. This is an excellent example of how Hollywood perceived Boyd as just another star in the period prior to the Hoppy television era.

Magazine Tear Sheet, *Saturday Evening Post*, Hopalong Cassidy Acme Cowboy Boots. The fact that these boots were not exact copies of those worn by Hoppy bothered no one. The only link with Hoppy is the identification tags used to help pull on the boots.

Sunday Newspaper Supplement, left half, "Make This Christmas A 'Hoppy' Holiday!," early 1950s. This Sunday supplement appeared in newspapers nationwide. It is an excellent for identifying clothing with specific manufacturers.

Sunday Newspaper Supplement, right half, "Make This Christmas A 'Hoppy' Holiday!," early 1950s. One of the hardest Hoppy items to find is a complete Barton Product's Hopalong Cassidy Slick-Up Kit. It retailed for $7.95, perhaps one of the reasons why parents purchased it far less than the Hoppy wrist watch at $4.95 or alarm clock at $2.95.

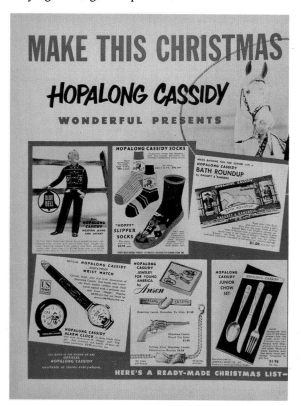

Catalog Tear Sheet, Montgomery Wards, 1951. Why buy the generic red cowgirl's outfit for $3.98, when, for the same price, you could own an official Hopalong Cassidy outfit? Hoppy outfits competed with a boys' space pilot's outfit, the frontier of the future in the 1950s.

Magazine Tear Sheet, Burnett's Instant Puddings, early 1950s. American Home Foods of New York held a license between March 1952 and April 1955 for puddings, pie fillings, and gelatin desserts. It was common for Hoppy food licenses to specify the exact recipe that would be used to make the product.

Recipe Booklet, Van Camp Sea Foods, Chicken of the Sea, six fold, 1953. Hoppy collectors seek this folder because one of its panel's includes "Hoppy Creed." Why did parent's love Hoppy? Rule 2: "Your parents are the best friends you have. Listen to them and obey their instructions." Nothing more need be said.

Recipe Booklet, Van Camp Sea Foods, Chicken of the Sea, six fold, 1953. Hoppy helped make tuna one of the most popular dishes of the mid-1950s. Many children's school lunch kits contained a tuna sandwich. Mine did.

Comic Book or Newspaper Tear Sheet, Post's Grape-Nut Flakes, early 1950s. Hopalong Cassidy products were advertised heavily in the Sunday comic pages and Hopalong Cassidy comic books. Archival records are minimal regarding the General Food Hopalong Cassidy licenses. The company does not appear on any of the licensing lists prepared by the Boyd staff.

Magazine Tear Sheet, Castle Films, early 1950s. This is another license product that appears to be handled by an outside agency, most likely Boyd's film distributor.

Magazine Tear Sheet, Aladdin Chuck Wagon School Lunch Kit and Vacuum Bottle. Aladdin did market the lunch kit and vacuum bottle as a unit and separately. The vacuum bottle sold separately for $1.69 and together with the lunch kit for $2.89.

Magazine Tear Sheet, Aladdin Chuck Wagon Kits and Lamps, early 1950s. While the survival rate of Aladdin lamps is relatively high, the shades are scarce. The presence of a shade can double the value of a lamp. Beware of sellers who have added a generic western shade in place of the Hoppy shade that came with the lamp.

Magazine Tear Sheet, Arvin' Hopalong Cassidy Radio, early 1950s. Arvin advertised the radio as having a crashproof case. Given the excellent case condition in which most radios are found, Arvin's claim appears to have had merit.

Catalog Tear Sheet, Sears, Roebuck, early 1950s. When collectors thinks of Sears cowboy merchandise, the name that most readily comes to mind is Roy Rogers. Many do not realize that Sears played the field. Sears sold tens of thousands of these Hopalong Cassidy wallets.

Catalog Tear Sheet, attributed to either Sears, Roebuck or Montgomery Ward, early 1950s. The Brewster Leather Goods Corporation of New York, NY, received a license in 1950 for children's school bags. Hassenfeld Brothers, Inc., of Pawtucket, Rhode Island, designed their products so that the same stamping die could be used for their album, scrapbook, and wastebasket.

Catalog Tear Sheet, Montgomery Wards' 1952 Christmas Catalog. One price, $7.65, serves all - Hoppy, Mickey Mouse, Alice in Wonderland, Cinderella, or Snow White. The original box triples the value of the watches.

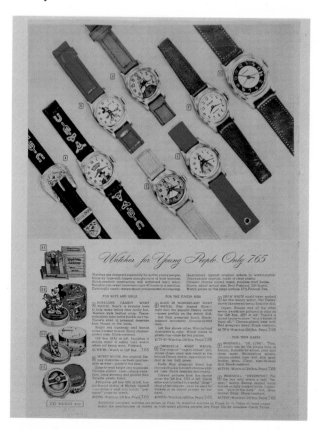

Premium Catalog, Hyde Park Dairies, Wichita, Kansas, 1956.
Several dairies used premium catalog promotions during the
mid-1950s. Each half gallon of milk was equivalent to two
points. It took 300 points to obtain the Hopalong Cassidy
pocket knife.

Magazine Tear Sheet, *Saturday Evening Post*, August 4, 1951.
Hoppy's the cowboy on the left. The Hoppy look-alike is
Hector DuBois of New York City.

Premium Catalog, Hyde Park Dairies, Wichita, Kansas, 1956. In
order for Hyde Park Dairies to attract young girls, they offered
dolls, a doll carriage, and a child's tea set, none of which were
licensed Hoppy products. The cowgirls of the early 1950s gave
way to young ladies cast from the traditional mold by the mid-
1950s.

Magazine Tear Sheet, unidentified origin, ca. 1949-50. Arvin
held the license for Hopalong Cassidy radios. Their venture
into television manufacture was short lived. There is no
mention of Hoppy or Boyd in the text of the advertisement.

Magazine Tear Sheet, *Saturday Evening Post*, February 18, 1950. Good-bye Arvin, hello Motorola. Hoppy is billed at the bottom as "America's favorite Television Cowboy."

Magazine Tear Sheet, *Life*. Once he held the character rights to Hopalong Cassidy, Boyd aggressively marketed the character. Nationwide tours were an important part of his marketing strategy. *Life* followed Hoppy as he greeted 35,000 fans in Oklahoma City and 50,000 fans in New Orleans.

Broadside. Boyd and his staff firmly believe in providing support material for their many commercial ventures. In addition to licensing several independent firms to produce this material, Hopalong Cassidy Enterprises (8907 Wilshire Boulevard, Beverly Hills, California), a firm Boyd owned, manufactured promotional materials for the media.

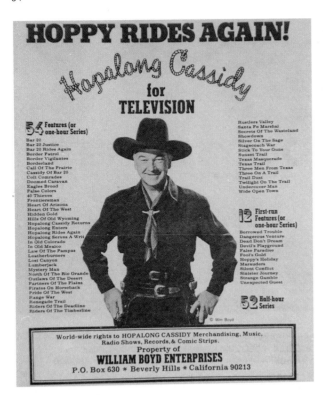

Magazine Tear Sheet, unidentified source. Numerous attempts, some meeting with limited success, were made to market the Hopalong Cassidy movies, television, and radio shows following Boyd's death in 1972.

Studio Promotional Still, late 1930s. The influence of Boyd's work in silent films is strongly evident in this pose. The smile and ease that is so much a part of Boyd's Hoppy photographs of the late 1940s and early 1950s is missing.

Notebook Filler, ca. 1940. Someone forgot to tell the colorist that Hoppy's neckerchief was blue and white. The vertebra slide remained a Hoppy trademark throughout his film and television career.

Bond Bread Promotional Card / Hopalong Cassidy Arcade Card. Contrast these two images with the two previous images. Hoppy has evolved into an approachable father figure, wise in the ways of the world. This image marks the blending of Hoppy's and Boyd's personalities as one.

Bank, Plastic, Hopalong Cassidy Savings Club. After making their initial deposit, savers were given a Hopalong Cassidy bank to encourage them to save more. The cooper tone example is much easier to find than the white. A blue version was sold through stores. **Reproduction alert.**

Promotional Photograph in Frame, early 1950s. The Decorative Art Corporation of Chicago used Hoppy's image to help sell its non-tarnish frames. Obviously, the company did not plan for its frames to last for forty plus years.

Birthday Card, Hopalong Cassidy Savings Club. With the exception of his dairy and bakery campaigns, Boyd's Hopalong Cassidy Savings Club was one of the most successful promotion that he licensed. Banks across America participated. **Reproduction alert**.

Pinback Button, Hopalong Cassidy Savings Club. A tiered savings plan was created. The more you saved, the higher you ranked. Initial investors were Tenderfeet. Ultimately, you could work your way up to Ranch Foreman.

Selection of Pinback Buttons, circa 1950s and later reproductions. The top row of buttons are from the early 1950s. The bottom row are 1980s/90s reproductions. It is hard to tell the difference from the front.

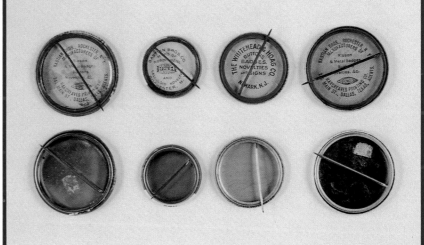

Selection of Pinback Buttons, circa 1950s and later reproductions. The difference between the period buttons and the later reproductions is obvious once you observe the back. Follow these rules to avoid being taken: (1) if it looks new, assume it is new and (2) if you have the slightest doubt, put it down and walk away. **Reproduction alert.**

Dressing Up with Hoppy

Over thirty-five companies were licensed to make Hopalong Cassidy clothing and clothing accessories. If it was worn, there was a Hoppy license for it. No aspect of daily life was left untouched.

After taking off your Hoppy pajamas by His Nibs Shirt Corporation, slippers by Acme Boot Company, and bathrobe by Van Baalen-Heilbrun and Company, it was time to put on your Hoppy underwear by Broadworth Mills and socks by Sport Wear Hosiery Mills. After adding a shirt by Little Champ of Hollywood and jeans by Blue Bell, don't forget to tighten your belt from Yale Belt Corporation. Pull on your boots by John A. Frye Shoe Company. Complete the outfit by adding a sweater by Barclay Knitwear Company, a neckerchief from Ernest Glick, and a hat from George S. Bailey Hat Company. If the weather was wet outside, you would put on your raincoat from Cable Raincoat Company, rubber boots from Goodyear Rubber Company, and gloves from Hugger Gloves. All forms of accessories from hair ribbons and tie clips to handkerchiefs and wallets were available.

Boyd never licensed the same product twice. He used a variety of methods to insure differences, e.g., quality of material, price point, single unit sale verse ensemble sale, boy's outfits verses girl's outfits, slipper verses sock slip-

per, etc. Whenever a product approached another in design, Boyd demanded a modification.

Boyd's overriding principle was value for one's money. While there was a Hoppy outfit available at every price point, Boyd did his best to insure it was the best buy possible. With a few exceptions, Hoppy clothing and accessories stood the tests of play and time exceedingly well. Because of its quality and durability, Hoppy clothing was often handed down. Even after two users, it held up remarkably well. Little wonder parents were willing to pay a slight premium to buy Hoppy clothing for their children.

Little clothing survives in its original packaging. It was meant to be used, and it was. As a result, collectors pay premium prices for clothing related boxes, e.g., the box in which Hopalong Cassidy socks were sent to the retailer (it had a colorful picture of Hoppy astride Topper) or the box from a pair of Acme Boot Company boots.

Although clothing related boxes are scarce, boxes for accessories are common. For whatever reason, accessory boxes were saved and used to store the pieces that came in them. If you encounter boxed material excellent condition, it probably is the result of a warehouse find or an unused gift. Several accessories, e.g., jewelry, were sold in sets. **Beware of sellers who break sets apart in an effort to increase their profits**.

Cowboy collectors (Hoppy collectors are no exception) visualize a 1950s child dressed exclusively in clothing and accessories related to their favorite television cowboy hero. Nothing is further from the truth. Children of the 1950s mixed and matched. So did their parents. It was The Look, not The *Authentic* Look that was important. One licensed piece was all that was necessary for a child to immerse themselves in the role of their favorite character.

Magazine Tear Sheet, *Life*, June 12, 1950. The caption for the photograph reads: "Hoppy Clothes from Lord & Taylor in New York, one of the many stores with 'Hopalong Cassidy Hitching Post,' shows a wide price range. From left, complete outfits include: with leather jacket ($45.20; with denim pants and shirt ($21.78); with girl's frontier set ($27.80); paneled shirt set ($29.30); cheapest girls' Hoppy outfit (hat $1.95 extra) at $4.95; suit with 'leisure jacket' ($42.15); and variations at $29.25 and $27.25. Prices vary very little throughout the country."

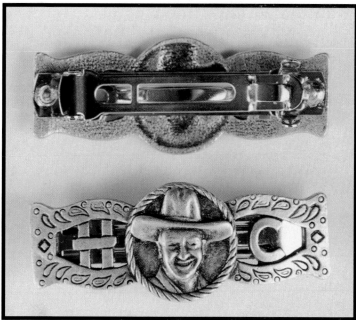

Barrette, Coro, Inc. Coro was one of the leading 1950s manufacturers of inexpensive women's costume jewelry. It held the Hoppy license for girls' jewelry and accessories.

Playsuit Box, Herman Iskin. Herman Iskin & Company, Souderton, Pennsylvania, received one of the most extensive Hoppy licenses. The company was licensed to make "boys' and girls' cowboy and cowgirl playsuit ensembles consisting of collapsible cotton hat, pants or skirt, vest, shirt, kerchief, lariat, belt holster or pistol; chaps and vest ensemble and skirt and vest ensemble; boys' boxed set ensemble consisting of pair of boxer-type longees, pair of boxer-type shorts and shirt."

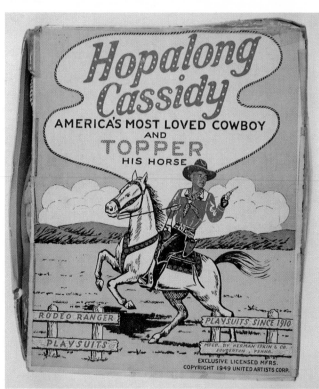

Hat, George S. Bailey Hat Company (Bailey of Hollywood). When buying the hat, make certain that the leather pull or slider on the chin string is present. It is necessary to make the hat complete.

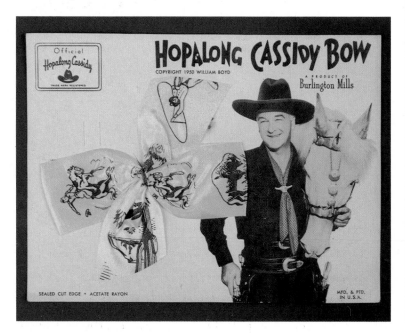

Bow, Burlington Mills. Burlington Mills Corporation held the license for girls' hair bow ribbon ensembles. This is an excellent example of how packaging adds value. The display-ability of the bow is enhanced by the Hoppy image on the card.

Playsuit, Herman Iskin. Boyd granted clothing manufacturers a great deal of flexibility in design and color. Not every outfit had to duplicate Hoppy's movie costume. Hoppy's identity was maintained through applied bust portrait images of Hoppy, Hoppy and Topper, and/or through text.

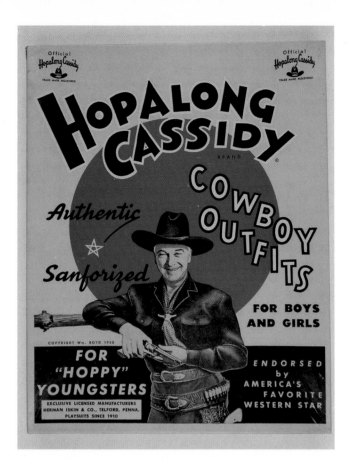

Playsuit skirt from Herman Eskin ensemble set. Note the loss of applied felt above the right horn of the steer. The skirt has cracked along several of its fold lines. These outfits are best left in the box and displayed as a boxed unit.

Playsuit Box, Herman Iskin. The large number of unused boxed sets of Herman Iskin's Hopalong Cassidy cowboy and cowgirl outfits is the result of a warehouse find in the early 1970s.

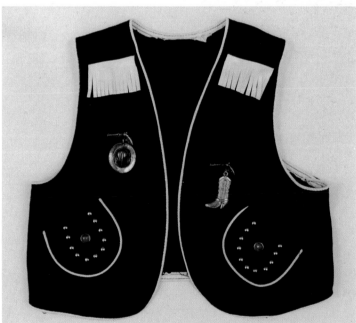

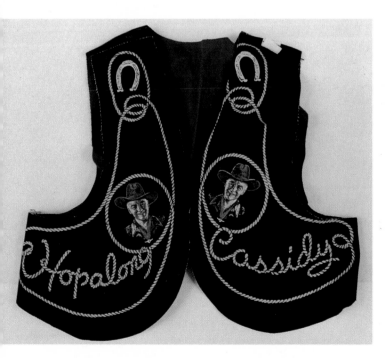

Vest, unidentified manufacturer. Unfortunately, the labels for this outfit have been lost. The lack of any specific Hoppy identification has caused some collectors to question whether or not this is an officially licensed product. The black and white color scheme suggests that it is.

Playsuit Vest from Herman Iskin ensemble set. This ensemble set was one of the least expensive Hopalong Cassidy outfits. As the ensemble ages, the fabric becomes increasingly brittle. Most cowgirl skirts have cracked on the folds. Preservation efforts are costly and largely non-effective.

Skirt, unidentified manufacturer. Hopalong Cassidy Enterprises, Inc., had an official identification and labeling program. Clothing manufactures bought the labels to be inserted in Hoppy licensed clothing from the Buffalo Woven Label Works. The company offered tag, woven, and printed label blanks.

Sweater, Barclay Knitwear Company. Barclay produced several types of sweaters. This is a short sleeve pullover. A long sleeve, button front and a sleeveless pullover are known.

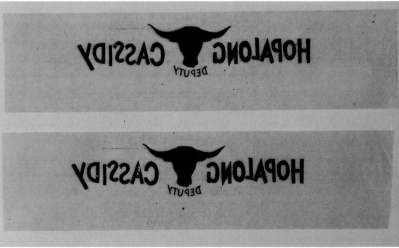

Decal, manufacturer unknown. Iron-on decals provided an inexpensive means to create one's own Hoppy outfit. Home sewing was extremely popular in the 1950s. Some of the "one of a kind" clothing that collectors encounter is more likely homemade than a factory sample.

Vest and chaps ensemble, unidentified manufacture. This is another example of a Hopalong Cassidy outfit that followed the black and white Hoppy theme. Note the lack of any specific Hoppy identification on the vest.

Shirt, Little Champ of Hollywood. Little Champ's license was for " boys' Western style shirts, sports shirts and cowboy shirts in rayon poplin and rayon gabardine." This was a strong middle market product.

Neckerchief and slide, Ernest Glick Company. In addition to neckwear, scarves, and handkerchiefs, Glick also manufactured Hoppy tie racks and brush holders.

Jacket, Richline. This is another questionable piece. While there is no Hoppy license for Richline, it could be a brand name for one of the licensed manufactures. The hat on the cowboy smacks of Hoppy. If it is a knockoff, it is an excellent one.

Jacket, Blue Bell. In the month of October 1950 Blue Bell reported wholesale sales of $425,064.84 included in which were 15,270 dozen of boys' cotton denim jeans. November was not a shabby month either - wholesales sales were $416,586.52, including 14,713 dozen boys' cotton denim jeans.

Hopalong Cassidy ensemble, mix-and-match. The level of Hoppy identification varied from outfit to outfit. The skirt contains the standard bust portrait of Hoppy and Topper surrounded by a lariat. The vest has minimal identification, the tie-in being the round conch with the black and white flaps

Cowboy ensemble, mix and match. What make's an outfit a Hoppy outfit? Clearly Boyd held no proprietary rights to the black and white color scheme. As a result, knockoffs were possible. This outfit with black and white felt plaid material on the shirt and inside the pants is one of the many licensed Herman Isken ensemble sets. The Hoppy identification is on the pants pocket. The gloves are generic, but very much in keeping with Hoppy's movie and television character. He wore gloves. What makes this a Hoppy outfit? The answer, of course, is the six-shooters.

Cowboy ensemble, mix and match. The denim jacket is a Hoppy licensed product. It is pictured in the June 12, 1950 *Life* magazines photograph of children wearing Hoppy clothes. The outfit underneath is from Herman Isken.

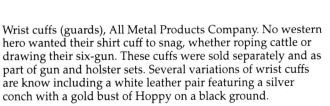

Wrist cuffs (guards), All Metal Products Company. No western hero wanted their shirt cuff to snag, whether roping cattle or drawing their six-gun. These cuffs were sold separately and as part of gun and holster sets. Several variations of wrist cuffs are know including a white leather pair featuring a silver conch with a gold bust of Hoppy on a black ground.

Cap Gun and Holster Set, All Metal Products Company. All Metal Products Company had a license to manufacture "single and double leather holster sets, cuffs, metal toys including cap pistols and spurs." All Metal's brand name was Wyndotte, the name normally used by collectors when talking about the guns. The "gold wash," double cap gun and holster set was the top of the line. Note the similarity in the decorative theme of the holsters with the wrist cuffs.

Cap Gun and Holster Set, All Metal Products Company. All Metal Products Company used a variety of holster styles to support the Hopalong Cassidy cap gun. The round conch with the steer head in the middle and a rope border is a convenient method of identifying an All Metal product.

Cap Gun and Holster Set, All Metal Products Company. Collectors use two characteristics - color and handle grip type - to differentiate the cap guns. The "silver" plastic handle guns were the least expensive. Note the enhancement of the steer conch with the addition of a leather strap.

Metal conch from All Metal Products Company holster. Eventually, All Metal Products added a Hoppy conch as a complement to its steer conch. Because these two adornments are so closely associated with a Hopalong Cassidy licensed products, individuals have been known to remove them from a badly damaged unit and apply them to generic cowboy material and faked items in order to pass them as *rare* examples of Hopalong Cassidy merchandise.

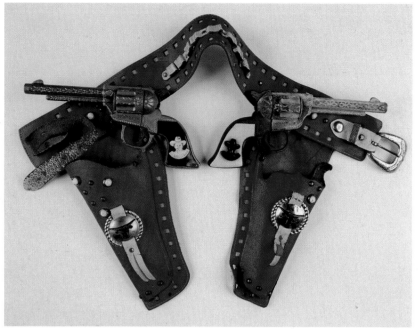

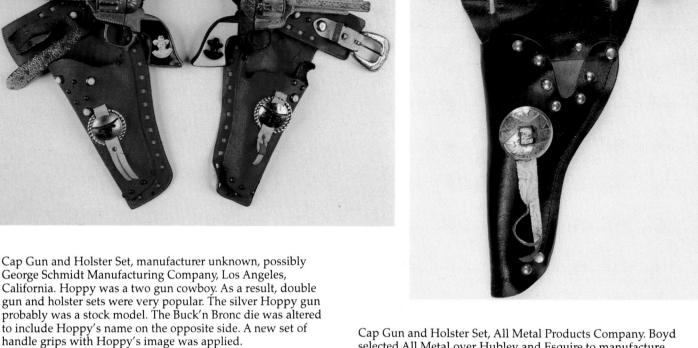

Cap Gun and Holster Set, manufacturer unknown, possibly George Schmidt Manufacturing Company, Los Angeles, California. Hoppy was a two gun cowboy. As a result, double gun and holster sets were very popular. The silver Hoppy gun probably was a stock model. The Buck'n Bronc die was altered to include Hoppy's name on the opposite side. A new set of handle grips with Hoppy's image was applied.

Cap Gun and Holster Set, All Metals Products Company. Collectors have speculated that the brown holsters were All Metal generic stock holsters and used in order to put a Hoppy cap gun set onto the shelves quickly plus provide time to develop a more thematic black and white (silver) holster. There is no evidence in the All Metal Products Company file to support or deny these assertions. Note the use of the Hoppy conch.

Cap Gun and Holster Set, All Metal Products Company. Boyd selected All Metal over Hubley and Esquire to manufacture Hoppy cap guns and holsters. All went well through the end of 1952. By October 1954, Hopalong Cassidy Co. had to send a letter to All Metals that stated: "Our contract with All Metal provides that they pay us a guarantee of $22,500 per year. In the contract year which ended September 30, 1954 their actual royalty payments were approximately only $7,500..." Note the conch variant on the holster.

Cap Gun and Holster Set, All Metals Products Company. By the time this set was issued, All Metal had created a completely new die for the Hopalong Cassidy cap pistol. These pistols have "HOPALONG" on both sides of the frame.

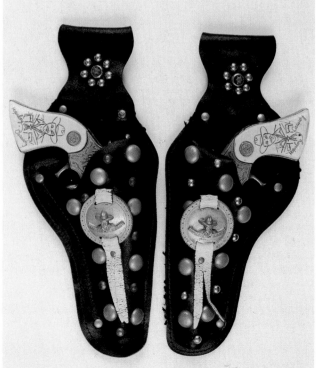

Belt Buckle, manufacturer unknown. Hoppy's trademark vertebrae neckerchief slide was converted into a steer's head motif and used with several Hoppy products including this belt buckle.

Holster Set, All Metals Products Company. One of the provisions of the All Metal contract was: "...it is understood and agreed that I reserve the right to license one other Holster manufacturer to manufacture, sell and distribute Holster Sets and guns in ensemble only, providing, however, such Holster manufacturer will purchase from you all guns required for such Holster Ensembles and providing further that such manufacturer is limited to sell such Holster and Gun in ensemble at retail prices which will not be less than approximately $5.98 per Holster set. You agree to sell such guns which may be required by said licensee at your normal wholesale price." A second license was issued to R & S. Toy Manufacturing Company, but not until April of 1957, long after the relationship with All Metal Products Company ended.

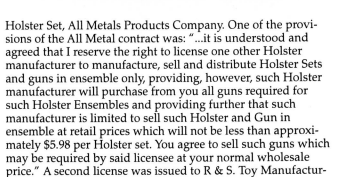

Belt, Yale Belt Corporation. Yale's license was for "children's leather belts, elastic suspenders, Leathercraft kits." The belt is Yale's Hopalong Cassidy Western Switch-A-Belt set. The belt has the "Frontier Trophy Buckle" on it.

Handkerchief, Ernest Glick Company. Glick produced a wide variety of Hoppy handkerchiefs with designs ranging from embroidered to printed. They often were sold in boxed sets. Glick's boxes featured a colorful drawing of Hoppy riding Topper and are now a collectible in their own right.

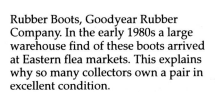

Rubber Boots, Goodyear Rubber Company. In the early 1980s a large warehouse find of these boots arrived at Eastern flea markets. This explains why so many collectors own a pair in excellent condition.

Rubber Boots, Goodyear Rubber Company. Contrast the differences between boots from the 1980s warehouse find and a pair that was worn. The rubber spur on the heel is usually missing. After a year or two of life in the mud puddle fast lane, the white on the boots discolored. If stored improperly, the rubber also becomes brittle and hard.

Children's socks, Sport-Wear Hosiery Mills. Continued washing damaged the screened Hoppy motif. Unwashed examples are difficult to find

Spur Set, All Metal Products Company All Metal used quality material. Their leather and metal usually survives the test of time well. Spurs are normally found in very good or better condition because of their limited use. Owners quickly tired of them. All Metal used the same box motif for the boxes for its pistols, holsters, and ensemble sets.

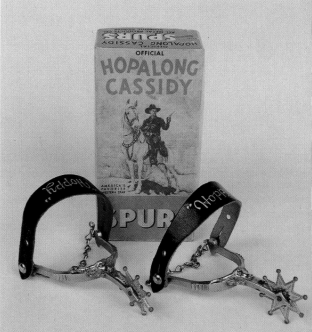

Child's Boot and Box, Acme Boot Manufacturing Company. Two separate contracts were issued to Acme — one for boots and another for shoes. The contract read "all types of shoes and slippers of whatsoever nature and description, bearing or using the name 'HOPALONG CASSIDY' and/or my likeness as 'HOPALONG CASSIDY.'" The company was excluded from making "rubber boots, rubber overshoes and rubber galoshes, and except children's hosiery, slipper sox, stockings." Acme sold 18,240 pairs of boots in September 1950. It does not require genius to see why the Acme Boot box is a desirable collectible in its own right.

Hoppy Fabric, manufacturer unidentified. Boyd licensed several different Hopalong Cassidy fabric patterns directed toward the home sewing market. Joseph Caro's *Collector's Guide to Hopalong Cassidy Memorabilia* (Gas City, IN: L-W Book Sales: 1992) pictures a child's bib-overall made from this fabric.

Wrist Watch, United States Time Corporation. While the bands on Hoppy wrist watches wore well, they did deteriorate over time. A popular 1950s replacement was the link, twist band. A collector will add this watch to his collection to show how individuals "valued" the Hoppy watch.

Wrist Watch, United States Time Corporation. United States Time had a license for "Children's pocket watches, wrist watches, alarm clocks." Several sizes of wrist watches were issued. The watch needs its period band to be considered complete. Two box variations are known, the most desirable being the saddle box. The box triples the value of the wrist watch.

Wrist Watches, United States Time Corporation. United States Time refused to sell parts to jewelers so their watches could be repaired. The company did its own repairs, charging fifty cents for the service in the early 1950s. Today a Hoppy watch can be repaired only by cannibalizing another watch or having a jeweler hand manufacture defective parts.

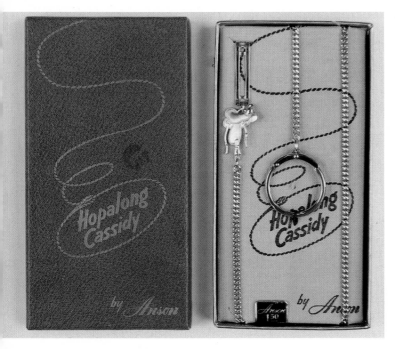

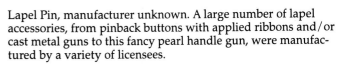

Lapel Pin, manufacturer unknown. A large number of lapel accessories, from pinback buttons with applied ribbons and/or cast metal guns to this fancy pearl handle gun, were manufactured by a variety of licensees.

Chain, Anson. Anson held the license for "boys' jewelry and accessories including chains, cuff links, money clips, and tie holders." In the 1950s, boys dressed as little men, especially for formal occasions. A tie was mandatory for Sunday church and dining out at a restaurant. As a result, a boy's jewelry chest included numerous tie bars, cuff links, and other assorted jewelry.

Tie Bar, Anson. Anson's $1.50 price point made its Hoppy jewelry very appealing to the parent or relative looking for a modestly priced Hoppy gift. Several warehouse finds of Anson material have been discovered when individuals bought the contents of old jewelry stores.

Lapel Pin, manufacturer unidentified. When the pin is opened, "HOPALONG CASSIDY" appears. Boyd licensed both his name and his image.

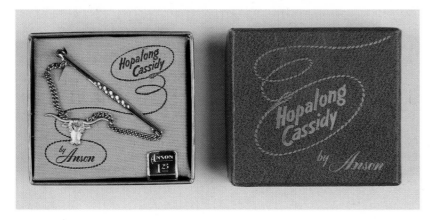

Wallet, Philip Florin Company. Florin manufactured close to a dozen different variations of Hoppy wallets, including one model sold exclusively by Sears. This example included a secret coin pouch with a Hoppy lucky coin. It is common to find the Hoppy lucky coin sold as a separate item at flea markets and collectibles shows.

Compass ring and two adjustable child's rings. The plastic Hoppy hat fits over the compass. The adjustable rings were inexpensive and sold primarily in five and dime stores.

Soap, Daggett & Ramsdell. In the 1950s cleanliness was next to godliness. No Hoppy cowboy or cowgirl would get dressed or go to bed without attending to their personal hygiene. Daggett & Ramsdell held the license for "children's soap, bubble bath, hair trainer, and shampoo." This four bar set had separate Hoppy decals applied to each bar. 400,000 bars of Hoppy soap are reported to have sold the first week it was offered for sale.

Wallet, Philip Florin Company. Florin held the license for "wallets, ring binders, spiral bound notebooks, utility kits, and fixed sets." Florin issued the wallets under its Pioneer brand name. Wallets came with a variety of inserts. The Hoppy Special Agents Pass was one of the most common.

Soap Box, Daggett & Ramsdell. One of the most desirable bars of soap to collectors is the three dimensional "Hopalong Cassidy's Horse 'Topper'" bar. As an added incentive the box contained a cut out Hoppy gun and targets. The lid of the box contained one side of the gun.

Dr. West Dental Kit, Weco Products Company. The complete Dr. West Dental Kit consists of a mirror with Hoppy's image on the surface, a Hoppy toothbrush in a plastic container with a Hoppy band, toothpaste in a tube featuring bust portraits of Hoppy, and a toothpaste box featuring a picture of Hoppy on Topper. The most common single item sold separately is the mirror.

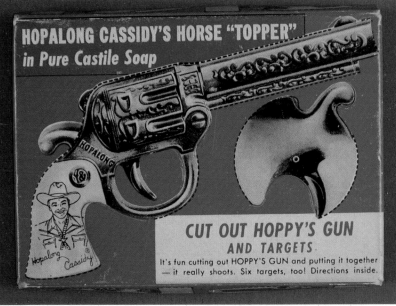

Dr. West Dental Kit Box, Weco Products Company. A large number of mint Hopalong Cassidy Dr. West's Dental Kit entered the collectors market in the early 1970s. In keeping with the premise that in today's market the sum of the parts is greater than the whole, dealers break the kits apart and sell the contents as separate units.

Soap Box, Daggett & Ramsdell. The base of the Topper box contained the other side of the cut out and the triggering mechanism. The six targets were on the side. Assembly instructions were inside the box. Because eager cowboys and cowgirls cut out the gun, few complete boxes are found.

Dr. West Dental Kit, Weco Products Company. This picture shows the initial sales presentation. The interior packaging cardboard is required to have a complete boxed unit.

Hair Trainer, Pint Bottle and 6 Ounce Bottle, Associated Brands. In an era when "a little dab'll do you," young cowboys used hair trainer on a daily basis. This is one of the few products where Boyd apparently issued multiple licenses. In additional to Associated Brands, hair trainer licenses were issued to Daggett & Ramsdell and Rubicon.

Hoppy at Meal Time

1950s cowboys and cowgirls had breakfast, lunch, dinner, and snacks with Hoppy. Once again nothing was overlooked, from the utensils used to eat the meal to the food products themselves. Since birthdays were always a special occasion, Boyd made certain that he offered a full line of products to allow him to star right beside with the birthday cowboy or cowgirl.

As always, Boyd was fussy about the products that he endorsed. Boyd's contract with Beatrice Food Company called for:

"A chocolate coated dairy bar which shall be not less than three ounces made in accordance with the following formula, modified as necessary to comply with all applicable Federal, State and Local laws:

Ingredients

3.0%	Butterfat - to comply with State requirements	
14.0%	Milk Solids - Not Fat	
13.0%	Cane Sugar	
4.0%	Dried Corn Syrup Solids	
.56%	Stabilizer	
34.56%	Total Solids	

on a stick and packaged in properly approved wrapper-bag, which package will bear the name HOPALONG CASSIDY and may bear the name and/or likeness of William Boyd in accordance with sample to be approved by us."

Boyd rejected a number of food product license applications after he tasted sample products and found them not to his liking.

Boyd apparently did not have full control over food licensing. The surviving records of Boyd's various licensing companies contain no information about cereal licensing and cereal premiums. There is no question that Boyd's Hoppy image was used extensively by Post cereals, especially on Grape-Nut Flakes and Raisin Bran. Raisin Bran issued a series of twelve lithographed tin badges as a Hoppy premium in the early 1950s.

Boyd's greatest meal time influence was dairy and bakery products. Boyd had a general contract with the Quality Bakers of America Cooperative that included over thirty five separate bakeries. His dairy industry licenses included over seventy-five licenses to dairies for milk, ice cream, and/or cottage cheese plus over twenty licenses limited solely to the manufacturer of ice cream. In both industries, Boyd controlled licensing by restricting geographic coverage. For example, the Sta-Kleen Bakery of Lynchburg, Virginia, was licensed to sell its products only in the Virginia counties of Campbell, Appomattox, Charlott, Prince Edward, Amherst, Bedford, Pittsylvania, Nattoway, Halifax, and Allegheny.

Since the individual bakeries and bakeries used essentially the same packaging and promotional products, Boyd licensed a series of manufacturers to make items sold only to these manufacturers. Marathon Paper Company (Menasha, Wisconsin), Pacific Waxed Paper Company (Seattle, Washington), Pollack Paper Corporation (Dallas,

Texas), and Western Waxed Paper Division of Crown Zellerback Corporation (Los Angeles, California) made end labels and bread labels. Milprint, Inc., Milwaukee, Wisconsin, made cellophane bread wrappings. Glass containers from milk bottles to tumblers came from Owens-Illinois Glass Company and Thatcher Glass Manufacturing Company.

Finding a food licensed product in its original packaging ranges from easy to difficult. Some, such as milk bottles, ice cream containers, and the Hopalong Cassidy Potato Chip can, are easy to locate. Bread wrappers, Betsy Ross grape juice bottles and cans, Burnett's Instant Pudding box, and Hopalong Cassidy Pop Corn cans are more difficult. Topps bubble gum card wrappers, Hoppy Candy Bar wrappers, Marlo Packing Company's canned beef stew and other products, and Lady's Choice Foods jelly jars fall into the most difficult category.

Place Mat, C. A. Reed Company. This is probably a placement from a Hopalong Cassidy Party set. Reed sold its Hoppy paper products in ensemble sets and individually. Dealers quickly destroy surviving Party sets and packs, preferring to sell the contents one piece at a time.

Plate, W. S. George Pottery Company. W. S. George had a license for "children's semi-vitreous dinnerware including plate, bowl, mug and/or cup." George used several decals for the center image of its plates.. When buying a set, make certain the decals mate.

Plate, W. S. George Pottery Company. Repeated washing faded the colors of the decal. Plates that retain their period luster command a premium. Also check carefully for knife cuts that have damaged the decal.

Mugs, W. S. George Company. The mug on the left is from the George "Chuck Wagon" set. The other may be contemporary fantasy item.

Cereal Bowl, W. S. George Company. In addition to selling pieces individually, W. S. George offered Hoppy china in a three piece set — dinner plate, cereal bowl, and mug. The "Chuck Wagon" set featured Hoppy in his classic black costume. A second set had Hoppy dressed in a blue outfit. The sets were sold in an attractive gift box. The box increases a set's value approximately $50.00.

Mugs, white milk glass, manufacturer unknown. Samuel Mallinger & Company of Pittsburgh, Pennsylvania, held a license for "children's glass cereal bowls, water and juice glasses, milk glass tumblers." While an attribution to Mallinger based on this license would appear appropriate, these glass mugs are believed to have started out life as peanut butter containers. If true, chances are they were made by another supplier.

Luncheon Plate and Mug, Milk Glass, Samuel Mallinger & Company. Mallinger offered a three piece, white milk glass, luncheon set - plate, bowl, and tumbler - that featured black bust portraits of Hoppy and/or Hoppy and Topper. Pieces from the Mallinger milk glass luncheon set are much harder to find than Hoppy's ceramic dinnerware.

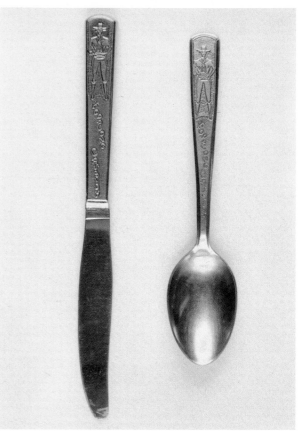

Glasses, obverse, Samuel Mallinger & Company. Mallinger made five, Hoppy, milk glass glasses - Breakfast Milk, Lunch Milk, Dinner Milk, Snack Milk, and Juice - in addition to the one included in the luncheon set. The front decal and inscription on the back of the glass are the keys to identification.

Knife and Spoon, Imperial Knife Company. Imperial Knife Company was licensed to manufacture "boys' pocket knives, juvenile stainless steel knife, fork and spoon set." Each implement has "HOPALONG CASSIDY" on the handle and a frontal view of Hoppy standing tall.

Glasses, reverse, Samuel Mallinger & Company. The reverse of the three glasses in the previous picture. The image and lettering on all the glasses is done with black ink. Mallinger sold a four glass boxed set that contained the breakfast, lunch, dinner, and snack glass. A plastic dinner glass, using the same art work as the milk glass dinner glass, has been documented. Additional plastic glasses may exist.

Fork, Imperial Knife Company. Imperial Knife marketed these utensils as the Hopalong Cassidy Junior Chow Set. The individual units are among the most commonly found Hopalong Cassidy items. However, boxed sets are difficult to find and a command premium prices.

Glass, obverse, manufacturer unknown. This glass is from a Hopalong Cassidy Western Series. While the glass set probably contained four to six different theme glasses,, thus far only I have recorded only Branding Iron, Indian Tribe, and Western Wagons glasses.

Glass, reverse, manufacturer unknown. The Hoppy image is far removed from those used by Mallinger; hence, the lack of an attribution to that company. The color scheme suggests a mid-1950s product, rather than one from the late 1940s or early 1950s.

Book Cover, Bond Bread. Bond Bread was a product of the General Baking Company of New York, New York. Hoppy helped it dominate the early 1950s store shelves in New York, New Jersey, and Pennsylvania. Bond Bread aggressively promoted its association with Hoppy. This book cover give-away is one of dozens of such items that Bond used.

Bread Package End Labels, General Backing Company. Bond's most successful promotion was a series of generic and numbered bread end labels. Children gobbled bread by the loaf in order to get Mom to buy more. The lucky child went to the store with his mother and carried a list of the numbers that he or she already had accumulated. Since the labels were exposed, the child could select a loaf with a label that he needed. Avid collectors found out when the weekly bread shipment was scheduled and were waiting at the store shelves when it arrived.

Hang-Up Album, Hoppy Saves the Gold Shipment, General Baking Company. Bond Bread created several story lines featuring its end labels. Children could obtain a Hang-Up Album in which to keep the labels they acquired and read the story in detail.

Hang-Up Album, Hoppy Saves the Gold Shipment, General Baking Company. This is the first story in the Bond Bread label series. In addition to providing a printed explanation for what is happening in each picture, the album also contains "Hopalong Cassidy's Creed for School Children," a third set of rules-to-live-by from Hoppy. The other two rules are Hoppy's Troopers Creed and the Hoppy Creed.

Hang-up Album, Hoppy Captures the Stagecoach Bandits, General Baking Company. This is the second story in the Bond Bread series. Each loaf of bread contained only one numbered end label. The other label was a generic promotional image. Stock boys love to frustrate young collectors by facing all the numbered label ends toward the interior of the bread shelves.

Bread Wrappers, Langendorf United Bakeries, San Francisco, California. Bread wrappers were printed in large rolls. Unused or partial rolls of several variations of Langendorf Butter-Nut Bread loaf wrappers surfaced in the early 1980s. Some examples were sold in strip form.

Hang-up Album, Hoppy Captures the Stagecoach Bandits, General Baking Company. Is there no end to Hoppy creeds? This album contains the "Hopalong Cassidy's Safety Creed." The third album, Hoppy Captures the Bank Robbers, touts "Hopalong Cassidy's Creed of Good Manners." Actually, Hoppy's common sense rules of life are one of the reasons why he enjoyed strong support from parents.

Bread Wrapper, Langendorf United Bakeries, San Francisco, California. Several enterprising dealers made cardboard inserts for the bread loafs and sold them as store displays, charging a premium price for the ordinary.

Cereal Premium, Post's Raisin Bran, early 1950s. The back of the lithograph tin badge identifies its origin. In addition to the badge series, Post also issued a set of thirty-six Western Collector cards, some of which contained Hoppy images and some of which did not.

Bread Wrapper, Langendorf United Bakeries, San Francisco, California. Langendorf received one of the largest territorial franchises. It was licensed for the entire states of Washington, Oregon, California, and the country of Washoe in Nevada.

Cereal Premium, Post's Raisin Bran, early 1950s. This lithograph tin badge is one of twelve badges in a series of "Hopalong Cassidy Badges," issued by Post in 195. It is the only badge in the series to feature Hoppy's image.

Cookie Jar, Peter Pan Products, Chicago, Illinois. Peter Pan took a stock western motif cookie jar and added a Hoppy label. None of the other motifs tie-in directly with Hoppy. The glue used to apply the label caused most of them to discolor. Loss of color on the decoration occurs because the color was applied above the glaze and comes off when rubbed with a harsh cloth.

Cookie Jar, Peter Pan Products, Chicago, Illinois. Once again, a stock cookie jar is made into a Hoppy product through the application of a decal. This example is missing its lid. The lid has embossed letters that read "Cookie Corral."

Container, Polk's Cottage Cheese, Polk Sanitary Milk Company, Indianapolis, Indiana. A surprisingly large number of ice cream and cottage cheese food containers have survived. Hoppy's devoted television followers also were avid savers. Never pay a premium for any Producer's container, glass or waxed cardboard. Remember, this company still licenses the Hopalong Cassidy image.

Ice Cream Containers, (left) one quart container from O'Fallon Quality Dairy, O'Fallon, Illinois, (middle) cup container, unidentified dairy, and (right) half gallon container from Purity Maid Products, New Albany, Indiana, early 1950s. O'Fallon Quality Dairy Company had only an ice cream license and its sales were restricted to Bond, Clinton, Madison, and St. Clair counties in Illinois. Purity Maid Products had a milk, ice cream, and cottage cheese license for Floyd, Harrison, and Washington counties in Indiana and Jefferson and Oldham counties in Kentucky.

Drinking Straws, Northwest Cone Company, Chicago, Illinois. Northwest Cone held the license for "ice cream cones, cups, packaged straws." Hopalong Cassidy drinking straws were sold under their Regal brand name.

Milk Bottle Cap, Linton Dairy, Wilmington, Ohio. Because all dairies utilized identical bottle cap lids, Boyd licensed one manufacturer to supply the industry. The supplier provided several generic designs as well as designs to which the dairy's name could be added.

Milk Bottles, mid-1950s. No matter what the dairy, the theme was always the same: "Hoppy's Favorite Milk." Quart bottles are relatively easy to find; pint bottles more difficult. Avoid buying any bottle whose printing is scratched or otherwise damaged.

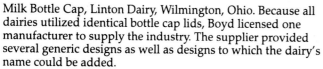

Milk Bottle Caps, (left) generic, (top) Dairyman's League Cooperative Association, New York, New York, and (right) Pevely Dairy Company, St. Louis, Missouri. The Dairyman's League held a milk, ice cream, and cottage cheese license for the entire state of New York, the northern tier counties of Pennsylvania, and the north portion of the state of New Jersey.

Milk Bottle, attributed to New York State. Most milk bottles were printed in a solid color with black being the most popular, followed by red. Collectors fill the bottles with dishing washing detergent to enhance their displayability.

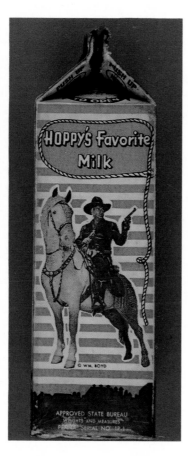

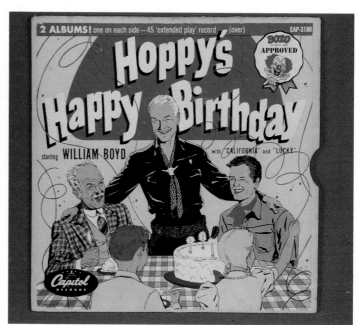

Record Album, Hoppy's Happy Birthday, Capitol. Capitol issued over a dozen album and single issue records in the 78 rpm and 45 rpm format. "Hoppy's Happy Birthday" was the ideal complement to C. A. Reed's Hopalong Cassidy Official Party Kit.

Milk Carton, Lehigh Valley Cooperative Farmers, Allentown, Pennsylvania. Waxed cartons are the hardest milk items to locate. They were not used by every dairy. The Hoppy image normally appears on the side rather than the front.

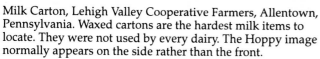

Invitation, Hopalong Cassidy Official Party Kit, C. A. Reed. Reed's Official Party Kit contained 12 invitations, 12 reply cards, a place mat game sheet, pin the tail on Topper, a Hoppy pin, and balloons imprinted with "HC."

Potato Chip Can, Gordon Foods, Atlanta, Georgia. Although a large number of potato chip cans survive, it is difficult to find one in fine or better condition. Most have slight to significant rusting and/or are dented and/or scratched. Deduct twenty to thirty percent for a missing lid.

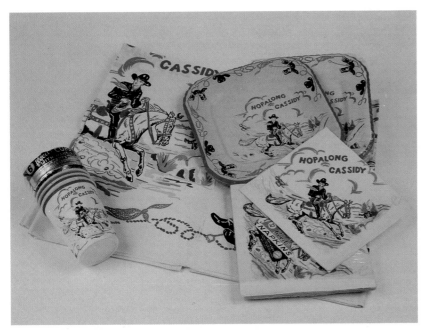

Paper Products, C. A. Reed. Among Reed's paper products with a Hoppy theme were dinner plates (package of 6), dessert/salad plates (package of 6), drinking cups (package of 6), napkins (package of 32), and a tablecloth.

Paper Napkin, C. A. Reed. Although redrawn for each product, this basic image tied the Reed paper product line together. When found in excellent condition, the colors are brilliant. Little wonder children found them attractive.

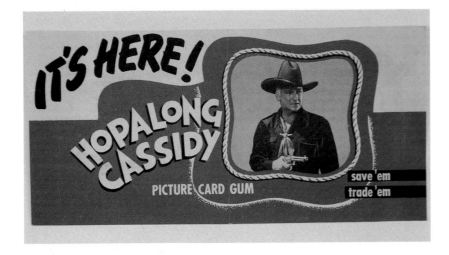

Broadside, Hopalong Cassidy Picture Gum Card, Topps. Topps released a 186, mono-chrome color, card series covering the story line of eight of the North Vine movies in 1950. The color changed for each new story. Two additional stories were covered by color cards (No. 187 through No. 230). Each of the eight monochrome stories had a foil header card. The gum cards were sold in penny and nickel packs.

Hoppy Goes to School

Boyd firmly believed in education - formal and informal. Some of the most pungent scenes in the Hopalong Cassidy movies are when Hoppy speaks with a young child about the future and the need for education. Of course, like all movie and television cowboy heroes of the 1940s and 50s, he defended the honor of the local schoolmarm on more than one occasion.

Many children wore western theme clothing to school in the early 1950s. If one examines the Hoppy clothing line carefully, one will see three basic types of outfits, those designed for (1) play, (2) school, and (3) dress-up occasions. Hoppy images and even the use of his name was minimized on items designed for the latter two uses. Manufacturers of other cowboy clothing lines, e.g., Gene Autry and Roy Rogers, followed the same approach. Of course,

the six-guns were checked at the door on the way out of the house.

Next to the lunch kit, the one Hoppy item that went to school on a daily basis was the Hoppy watch. Children wore watches; and, a Hoppy watch was a status symbol. Unlike the inexpensive dime store rings and other cumbersome jewelry, the Hoppy watch had class.

Hoppy school items fall into four classifications: (1) lunch kits, (2) stationary and writing supplies, (3) miscellaneous paper items, and (4) bicycles. Objects such as coloring and crayon sets more correctly belong in the "At Play" category. In addition to school items designed to be sold through normal marketing channels, several giveaways, e.g., book covers, also were school oriented.

A number of Hoppy items that probably should have been left at home were snuck into school. While Imperial Knife's large Hoppy pocket knife was easy to detect, most teachers failed to find the smaller version hidden in many a boy's pocket. Hoppy bubble gum cards were traded at recess as eagerly as baseball cards. More than one Hopalong Cassidy comic book was hidden among the papers in a Hoppy school bag and read during times when the student should have been engaged in more constructive activities.

Lunch Kits, Aladdin Industries, Nashville, Tennessee. The first Aladdin Hoppy lunch kits reached the store shelves in May 1950. Imagine Aladdin's delight when sales went from $295.04 in May to $130,903.01 in July and to $201,065.82 in August. Joy turned to sorrow years later when Boyd enforced a contract clause that read: " You further agree that you will not manufacturer, sell or distribute any of the articles specified in Paragraph 22 hereof or any article directly similar thereto which is identified with or bears the name of any living or dead person of prominence or any cowboy personality." Aladdin watched Gene, Roy, and others turn elsewhere, providing the opportunity for rival companies to develop and challenge Aladdin's market position.

Vacuum Bottles, Aladdin Industries. Although a collector today refers to these objects as thermos bottles, in 1950 Thermos was a major rival of Aladdin in the manufacture of vacuum bottles. While a change in color or design may not represent a significant change to a manufacturer, it does to a collector. Note the three vacuum bottle top variants. No collection is complete without one of each.

Lunch Kit, Aladdin Industries. In 1954 Aladdin introduced a Hoppy lithographed western scene lunch box as a means of stimulating sales. The vacuum bottle also had a new decorative motif. Aladdin also asked to have its marketing territory increased to include Mexico, Guatemala, El Salvador, Cuba, Honduras, Nicaragua, Costa Rica, Columbia, Venezuela, Netherlands West Indies, Haiti, Dominican Republic, Philippine Islands, and Siam. Little further proof is needed of Hoppy's worldwide popularity.

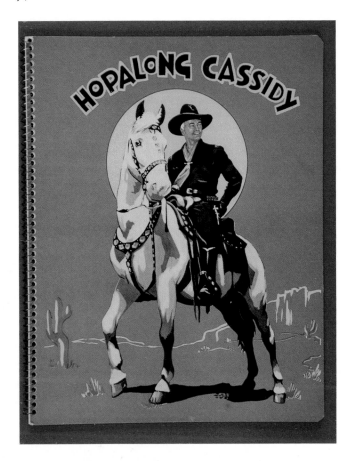

Spiral Notebook, Philip Florin Company. The art work on this notebook is among the best available on a mass-produced product. It captures the essence of the television Hoppy. Florin used the same art work on its two ring binder.

Tablet, manufacturer unknown, possibly Hassenfeld. This is the same picture that was used on the previous tablet. The principal difference is that a solid green color has been used to mask out the background. Beware of sellers who tear the cover off the table and attempt to sell it as an unusual promotional picture. The brown spots on Topper are a dead giveaway as to its origin.

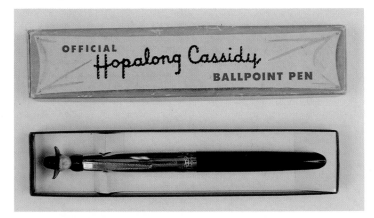

Ball-point Pen, Parker Pen Company, Janesville, Wisconsin. Parker made a fountain and ball-point Hoppy pen. The ball-point version is much easier to find than the fountain version. Boyd's commitment to quality is evident in his Parker license. Parker was one of the top pen manufacturers of the 1950s.

Tablet, manufacturer unidentified, possibly Hassenfeld. Inexpensive school tablets were readily available at five and dime stores. The person who did the hand colored overlays for printing mistook shadows for brown spots on Topper.

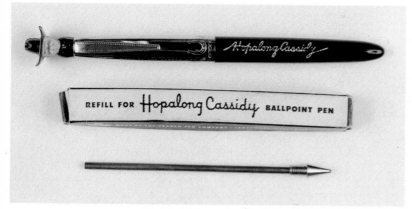

Ball-point Pen and refill, Parker Pen Company. Refill units in their original box are common, not scarce. As this book was being written a new hoard of twenty plus examples is working its way into the market.

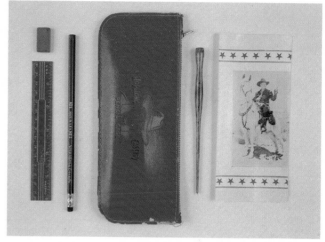

Zippered Pencil Pouch, Hassenfeld. The picture, pouch, and pencil have Hoppy identification. The pen nib holder, ruler, and eraser are generic. This pouch is obviously targeted toward the lower end of the market.

Pencil Light, battery operated, manufacturer unknown. When the eraser is turned, a light shines through the bottom plastic portion revealing a bust portrait of Hoppy.

Pencil Case, attributed to Hassenfeld. The Hassenfeld license was very specific. The company could make "plain and zippered pencil pouches, pencils encased in wood (except mechanical pencils), erasers, rulers, pencil sharpeners, and school children's pencil sets in paper, artificial leather, leather, and plastic pencil boxes." Is cardboard the same as paper? If yes, this is a Hassenfeld product. Chances are that it is.

Pencil Case, Hassenfeld Brothers. This was designed to fit into a shirt pocket or school bag. It contained a plastic pen nib holder, #2 pencil with "Hopalong Cassidy" on its side, and a clear plastic ruler.

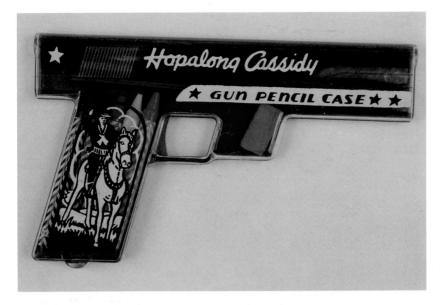

Gun Pencil Case, Hassenfeld Brothers. Designed for the younger crowd, the contents consist of five crayons, an eraser, plastic pen nib holder, and green plastic ruler. If Boyd were alive today, I feel certain he would never license a product that suggested taking a gun to school. How times have changed!

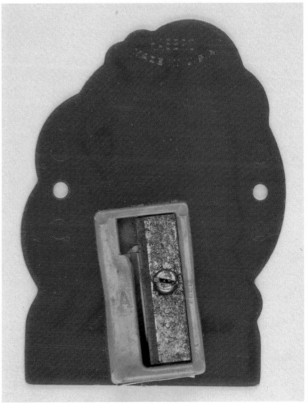

Gun Pencil Case, Hassenfeld Brothers. It is not clear if the three pencils included in this case are period. Most other Hassenfeld Hoppy products have pencils imprinted with "Hopalong Cassidy." The crayon coverings do match those in the gun pencil case pictured previously. While the holster was a nice touch, one has to question how positive a reception it received from the teacher.

Pencil Sharpener, Hassenfeld Brothers. Hassenfeld used the same art from its pencil gun case holder for its pencil sharpener. The back of the pencil holder provides a clue to Hassenfeld Brothers modern identify. Few collectors would fail to recognize the name: HASBRO.

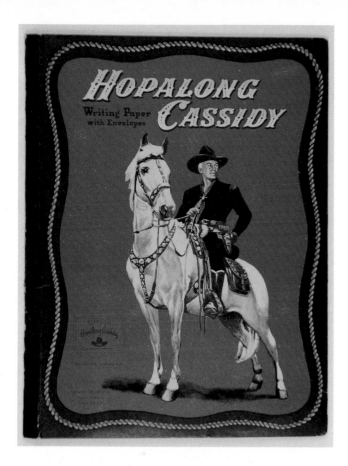

Writing Paper and Envelopes, Western Printing and Lithography Company. Using its Whitman Publishing Company division, Western Printing and Lithograph Company, signed a license for "jigsaw puzzles, boxed Sticker Fun cutout books and cutouts, children's stationery." The paper and envelopes in the set contained a bust portrait of Hoppy in the upper left corner.

Note Card, Buzza Cardozo Corporation. Buzza Cardozo held a license for "greeting cards, party invitations, notes, and birth announcements." Cardozo added variety to their cards through the use of balloons, die cuts, foldouts, game board and spinner, pop-ups, and other types of novelties inserts.

Note Card, Buzza Cardozo Corporation. Most cards worked on a four fold principle. Note the use of a Hoppy pin as a surprise gift. This is the same pin that was sold with a ribbon and attachment, e.g., white metal gun.

School Bag, Brewster Leather Goods Corporation, New York, New York. Brewster was licensed to produce "children's Western style handled school bag." Chances were better than even that another person in your class owned an identical Hoppy item. How do you tell your school bag from someone else's? Write your name on the front in ink. This personalization, while important in interpreting the school bag within its social context, is considered a negative by collectors. Deduct twenty to thirty percent. Do not pass go.

Bicycle, girl's model, D. P. Harris Hardware Manufacturing Company, New York, New York. A Hopalong Cassidy bicycle is considered one of the premier pieces in any collection. Harris marketed the bikes under its Rollfast tradename. Because of the demand for these bicycle, some individuals have repainted other Rollfast models and tried to pass them off as Hoppy bikes. In the early 1990s Hammer Schlammer, a major mail order catalog, commissioned the restoration of several damaged bikes and sold them at premium prices.

Hoppy at Play

The Fifties child had a great deal of leisure time. Over half the homes did not have a television. Further, parents carefully controlled children's viewing hours in those that did. No one was quite certain whether television was a positive or negative influence. Hoppy's strong emphasis on moral values and right and wrong helped convince doubting parents that it was

Children played games - outside when the weather was good and inside when it was bad. Outside children's imaginations ran wild in the 1950s. Cowboys and Indians, war, and space were popular topics for outdoor hide and seek, shoot'em up games. While everyone lived to fight another day, it was quite common for a child to be killed several times during an afternoon.

Life was relatively simplistic. The Cold War reduced everything to good verses evil, black verses white. Some how Hoppy stood for all the right things at the right time. The social relevancy of the Sixties was five to ten years in the future.

The rough and tumble nature of outdoor play required objects that could "take a licking and keep on ticking." Boyd licensed products that passed and exceeded that test. This is why so many survive in such good condition.

A common method of passing time in the evening, once chores and school work were done, was to play a game. Most were competitive in nature with a clearly defined winner and loser. Entire families participated; hence, the emphasis on games that were challenging enough to hold the interest of an adult, but not too complex to confuse a youngster. Canasta was the card game of choice.

Hoppy products designed for play are largely licensed versions of generic games with a long established track record, e.g., Chinese checkers and dominos. The Fifties were not a period of great innovations in the toy, game, and puzzle community. Everyone played it safe, sticking with the tried and true. A few new products, e.g., an inflatable rubber Topper life ring for use in a swimming pool and the Hoppy television set, did arise in response to the demands of the suburbia movement or as a reflection of the growing importance of television.

Tricycle, D. P. Harris Hardware Manufacturing Company. Harris' license was for "children's bicycles, velocipedes, tribikes, sidewalk roller skates, bicycle saddle bags." Tricycles are much harder to find than bicycles. Note the use as a decorative element of the same Hoppy conch found on All Metal's holster sets.

1950 to 1953 was the Golden Age of Hoppy at Play. Many manufacturers left the playing field by the 1954 Christmas season. Except for a few select items, the sale of Hoppy merchandise was seasonal. The Christmas season (much shortened in the 1950s) followed by the back-to-school season were the two best periods.

Further, it is important to look at the broad picture. While Hoppy dominated the scene from 1949 to 1951, Gene and Roy were riding hard to catch-up. Hoppy started with such a large share of the market that the only way to go was down. As more and more television cowboys stars and programs jumped aboard the licensing bandwagon, Hoppy's market share continued to shrink. By 1954 he was just another face in the crowd. As the Fifties ended, most of the children that grew up with Hoppy had put their toys into storage. An era as well as a decade had ended.

Canasta Ensemble, Pacific Playing Card Company, Los Angeles, California. The saddle motif proved ideal for canasta. In addition to being sold as a set, the saddle holder and cards could be bought separately.

Binoculars, Galter Products Company, Chicago, Illinois. Galter held the license for Hopalong Cassidy binoculars and camera. This is a generic child's binocular to which a Hopalong Cassidy decal has been added. In 1953 the Galter license was assigned to the Herold Manufacturing Company, Chicago. Binoculars with a gray ground and red eye-piece are a variant.

Camera, Galter Products Company. This is a basic plastic box camera with a screened Hoppy plate on the front. A flash attachment with a silk screened image of Hoppy on Topper on the inside of the reflector was available as an extra. Galter also made a 35mm, plastic body camera that had Hoppy's name on the front ring. The cameras came in boxes with strong graphics.

Canasta Playing Cards, Pacific Playing Card Company. Cards featuring Western motifs instead of numbers were used. No attempt was made to have the images portray other licensed Hoppy products.

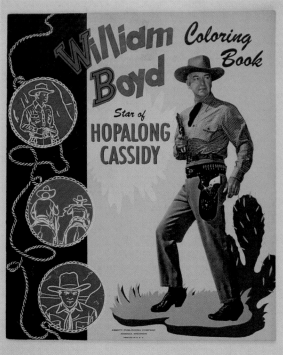

Chinese Checkers, Milton Bradley Company. Is it mere coincidence or the work of a higher power that the playing field for Chinese checkers is identical to the typical star shaped sheriff's badge? Milton Bradley introduced the Hoppy Chinese Checkers game in 1951 and sold a total of 74,038 units over the five years that it was offered for sale.

Coloring Book, Lowe, Inc., Kenosha, Wisconsin. Lowe, Inc.'s Abbott Publishing Company published this William Boyd Coloring Book. Throughout the mid-1950s Boyd made several attempts to capitalize on his own persona, most notably in the comic book area. Success was minor.

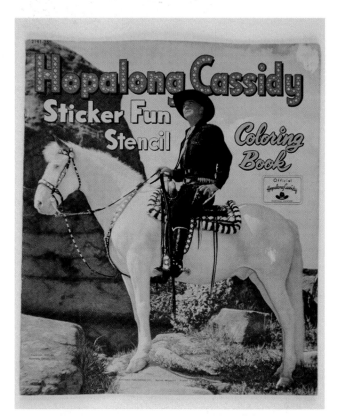

Sticker Fun / Stencil / Coloring Book, Western Printing and Lithography Company. Western's Whitman Publishing Company avoided infringing on other publisher's coloring book licenses by combining stickers with coloring. Deduct forty to fifty percent from the value of any coloring book that has been used, not matter how well the individual stayed within the lines.

Coloring Book, Lowe, Inc. A typical Hoppy coloring book is thirty-two pages in length. Some have a story line, others do not. On occasions new covers were done to give an old coloring book a fresh marketing appearance.

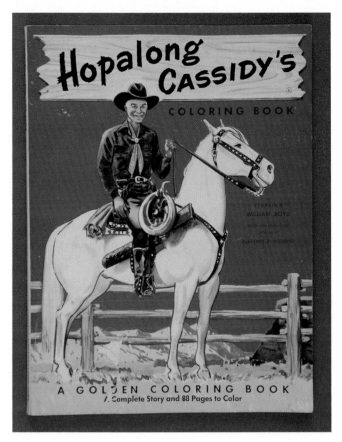

Coloring Outfit, Transogram Company, 1950. Transogram Company, a major game manufacturer in the 1950s, received a license for "crayon and paint sets, Ring-toss game, horseshoe pitching sets, bean bags, slate, toy pastry cooking pans, television chairs." Coupled with the license issued to Milton Bradley, these two firms dominated the Hoppy game market.

Coloring Book, Western Printing and Lithography Company. Western worked closely with Simon and Schuster in the development of Little Golden Books. The eighty-eight page length and strong story line suggests a licensing arrangement via Doubleday as opposed to Boyd.

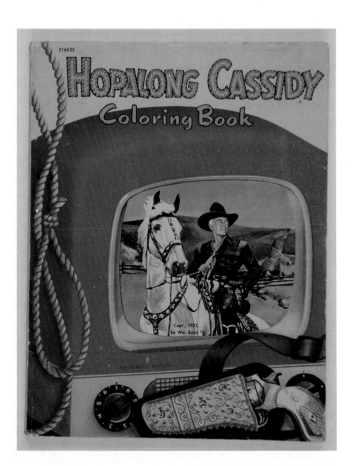

Coloring Outfit, Transogram Company. Transogram did an excellent job on the design of this Western motif coloring set. Hoppy's image appears prominently throughout the set. Carefully note the bottles and their caps. These often are removed from a broken set and sold as Hopalong Cassidy ink bottles. Boyd never licensed an ink product.

Coloring Book, manufacturer unknown, possibly English. The art work strongly suggest an English origin since it matches the covers of the Adprint annuals, but the spelling is American. Further, a check of foreign licenses reveals no licenses were issued for coloring books.

Wrist Compass, manufacturer unknown. There are no markings of any kind on the compass housing or band. The compass has a translucent, milky band with cowboy motifs. A later Japanese knockoff copies the band and compass housing exactly, but substitutes a watch in place of the compass.

Crayon and Stencil Set, Transogram. Transogram used the same stencil patterns in this game as it did for its Coloring Set. A row of jumbo crayons is located underneath the stencils. Some sellers have stripped the crayons from the set and offered them for sale as a Hopalong Cassidy crayon set. As separate boxed crayon set was not licensed.

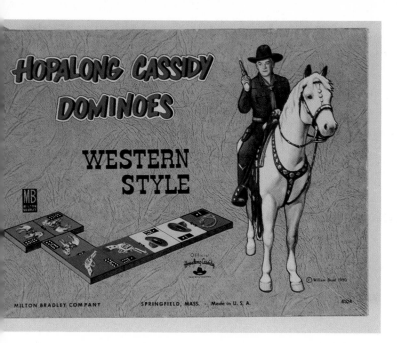

Dominoes, Western Style, Milton Bradley Company. When Milton Bradley introduced the set in 1951, it sold 29,860 units. Sales dropped to 25,380 units in 1952, 11,309 units in 1953, and 175 units in 1974.

Figurine, plastic, Ideal Toy and Novelty Company. This is an action toy. The hand holding the beaded reins can be moved to lift up and grasp the corner of the hat. The cowboy then appears to be waving the hat over his head.

Game, Boxed Board, Milton Bradley. The Hopalong Cassidy Game was a big seller. Over 400,000 units were sold in six years. A minor change was made in the game in 1953 as a means of increasing sales. The effort worked to some degree, doubling the amount of 1952 sales in 1953 and 1954. The game contains a large number of parts. Check the instructions and count the parts carefully before paying the price a complete game commands.

Doll, Ideal Toy and Novelty Company. Boyd licensed Ideal to produce "dolls styled as cowboys, horses (some mechanical), plastic rifle with metal parts." The only Hoppy identification for this doll is the hat band and label on its shipping box. Most dolls are missing the plastic holsters that attach to the belt and the two white metal guns that they held. Because of its generic nature, this doll is not high on the list of Hoppy collectibles.

Catalog Tear Sheet, Flashlight advertisement. The advertisement stresses three features: (1) regular 400 foot beam, (2) flasher button, and (3) red plastic head ring for danger signals. Stressed is "Attractively boxed to create real excitement on Christmas morning!" The flashlight proved to be a popular stocking stuffer at 98¢.

Figurine, plastic, Hartland. This is not a Boyd licensed product although collectors and sellers often confuse it for one. While generic, it is obvious that Hartland capitalized on the Hopalong Cassidy craze to market its product.

Flashlight, H. J. Ashe Company. This flashlight had the Morse Code on its back. The top unscrews to add the batteries. A later version had a siren attached to the end. Both flashlights were packaged in highly graphic boxes.

Gym Set, Dyer Products Company. One of the more unusual of the Hopalong Cassidy collectibles. Strangely enough, complete units are more easily found than individual parts. For whatever reasons, this is one toy children stored as a unit in its period box.

Lasso Game, Transogram. Highly desirable to collectors because of its strong displayability factor. A complete game should have six rope rings. The box contains a strong graphic of roping cattle near a corral.

Picture Gun Theater, Stephens Products Company. In order to construct the theater, the box itself was used. The bottom became the base. The insert steadied the wall panels. The lid fit on the wall panels to complete the transition.

Picture Gun Theater, Stephens Products Company, Middletown, Connecticut. Many families had 8mm home movie projectors. The Picture Gun Theater allowed children to copy their parents. A battery powers a bulb that projects light through the front of the gun. The film is loaded into the gun and advanced a frame at a time by pulling the trigger.

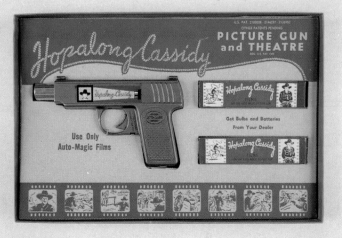

Picture Gun Theater, Stephens Products Company. This toy is an excellent example of a company taking an existing product and adapting it for a Hoppy license. Picture Gun toys existed for several licenses characters. Hoppy was not the first.

Pocket (Trail) Knife, Imperial Knife Company, Providence, Rhode Island. One of the greatest Hoppy licensing successes. Over a million units were sold the first week. Imperial reported total sales of $157,832.61 in May 1950. Boyd's royalties for that one month alone were $7,891.63. Most knifes have lost their plastic belt loop.

Puppet, Hand, National Mask & Puppet Corporation, Brooklyn, New York. National Mask and Puppet had a license for hand and string puppet and a toy chest hassock. The company also made a full image hand puppet with the head, hands, and boots made of rubber. Often missing is the white metal gun for the puppet's holster. No example of the toy chest hassock is known.

Pocket Knife, Imperial Knife Company. In addition to its Trail Knife, Imperial made a miniature pocket knife with only one blade. These are much scarcer than their larger brother.

Record Jacket, Billy and the Bandit, 45 rpm, Capitol. Capitol issued over a dozen single records with a Hopalong Cassidy theme. Billy and the Bandit is in a series that includes a 78 rpm entitled "My Horse Topper."

Jigsaw Puzzle Set No. 4025, Milton Bradley Company. This three puzzle set was Milton Bradley's biggest seller. It sold 578,486 units in the three years of 1950 through 1952. An additional 57,352 units of a modified example were sold between 1953 and 1955. These puzzles were never sold separately. Refuse to buy any puzzle marked 4025x1, 4025x2, or 4025x3, when offered for sale as a single unit. A single unit means a greedy seller has broken the set apart as a means of gouging the buyer.

Jigsaw Puzzle Set No. 4103, Milton Bradley Company. This puzzle set is a direct attempt to capitalize on Hoppy's television popularity. While the photographs appear to be from a stock house, they do not appear on any other Hoppy licensed product. This puzzle set was expensive by 1950s standards which may explain by only 60,393 total units were sold.

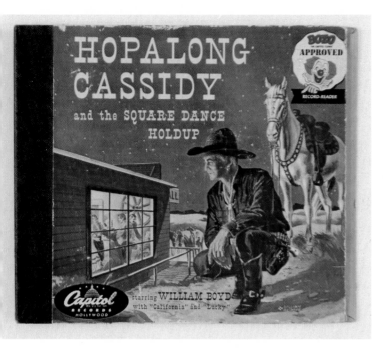

Record Album, Hopalong Cassidy and the Square Dance Holdup, Capitol. This and its companion album, Hopalong Cassidy and the Singing Bandit, were issued in 78 rpm and 45 rpm versions. They were Capitols most successful Hoppy records. The albums include a story book. Children turned the page when Topper whinnied.

Record Album, Hopalong Cassidy and The Story of Topper, 45 rpm, Capitol. The real collectible value of Hoppy records rests in their cover art, not in the records. Most owners never even play them.

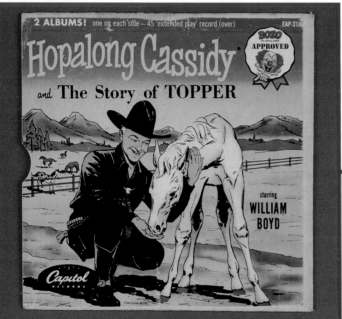

Shooting Gallery, Automatic Toy Company, Staten Island, New York. A spring action gun is used to fire metal bearings at the moving targets. The original box doubles the value of the game.

Rocking Horse, Rich Industries, Clinton, Iowa. The rocking horse has a plastic body and wooden legs and rockers.

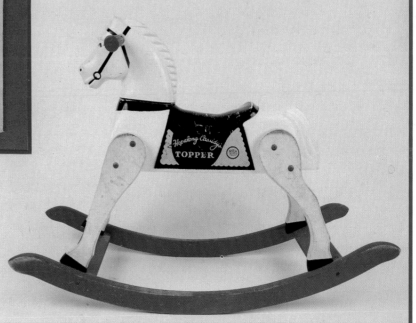

Inflatable Vinyl Plastic Topper Flotation Toy, U.S. Fiber & Plastics Corporation, Stirling, New Jersey. This toy was designed for use in bathtubs and pools. Boyd licensed two other companies to make inflatable toys: (1) Aubin-Lee Rubber Products Corporation of California, Gardena, California and its successor companies — Lee-Tex Rubber Products of California and Lee-Tex Rubber Products of Illinois and (2) Urb Plastic Company of New York, New York. Urb's license was for "bouncing type toys vinyl plastic inflated Hobby Horse with automatic noise maker and wading pool."

Roller Skates, D. P. Harris Hardware Manufacturing Company. Like the Hoppy Bike, these roller skates were marketed under the Rollfast tradename. The skates had spurs that attached and wheels with gold colored hubs. The skates were manufactured at Harris' plant in Reading, Pennsylvania.

Shooting Target Game, Louis Marx, New York, New York. Marx was a major manufacturer of lithograph tin toys and games in the 1950s. This target is part of a set that includes a dart gun and rubber tipped darts.

Target Game, Toy Enterprises of America. The game was marketed as two games in one. This is the reverse of the previous illustration. The cardboard box for the target game utilized the same images as on the game's surface.

Mechanical, Shooting Target Game, Louis Marx. As more and more collectors research the origins of Hopalong Cassidy products, mechanicals and other production materials are being discovered in the hands of former factory workers or company archives, saved primarily for the reason that they were too important to throw out.

Target Game, Toy Enterprises of America, Chicago, Illinois. Toy Enterprises had a license to make a "Magne-Safe Arrow magnetic dart game." Note that Topper and Hoppy's outfit is hand drawn. Only Boyd's face is from a photograph.

Television Set, Automatic Toy Company. The television set was key wound. As sense of movement was achieved as the roll passed in front of the bars. The set had four separate rolls plus two bonus promotional pictures.

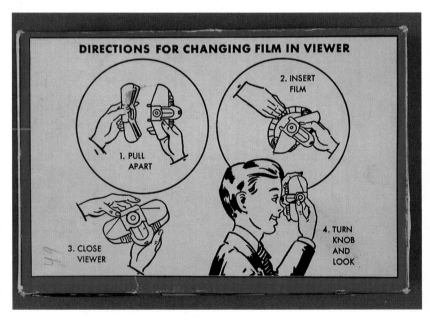

Movie Viewer, manufacturer unknown. This inexpensive film viewer was sold in a variety of packaging from a box set as show to plastic packages designed for rack sales. Extra films could be purchased in packs.

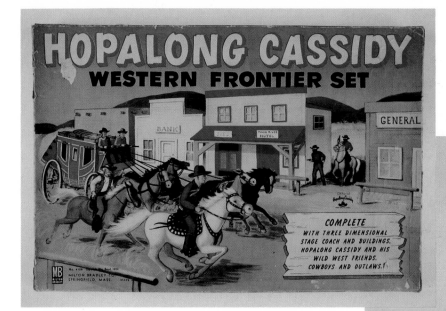

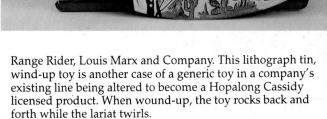

Western Frontier Set, No. 4108, Milton Bradley Company. This set allowed a child to recreate a Western town on a table top. Assembly instructions are located on the underside of the box lid. Milton Bradley sold 30,008 units over a four year period, with 17,305 units being sold in 1951, the first year of issue.

Range Rider, Louis Marx and Company. This lithograph tin, wind-up toy is another case of a generic toy in a company's existing line being altered to become a Hopalong Cassidy licensed product. When wound-up, the toy rocks back and forth while the lariat twirls.

Wood Burning Set, American Toy & Furniture Company, Chicago, Illinois. The popularity of wood burning kits was nearing its end in the early 1950s. This type of toy was extremely popular in the late 1930s and underwent a brief revival in the late 1940s. Less than 10,000 sets were sold in 1950. Just over 27,000 units were sold in 1951. Sales went totally flat in 1952.

Hoppy's Bunk House

The one thing you could not do was furnish your entire house in Hoppy. Apparently the parents of the Fifties were perfectly willing to tolerate the Western look in the bedroom and den, but not in the kitchen, living room, or elsewhere in the house. Thank goodness. The mere thought of a Western style, red Formica top table is enough to boggle even my wildest imagination.

Boyd's principal licensing effort in respect to furniture and furniture accessories was the bedroom, the Fifties child's equivalent of the bunk house. The Western bunk house was a place to rest, relax, and sleep. Meals were served there. However, preparing and serving them always was portrayed as a secondary activity in Western movies and television shows. The bunk house was a place to let your hair down, the location for camaraderie, congenial times, and horsing around. Many a Western tune was played and sung within the confines of the bunk house walls.

Boyd product licensing was so thorough that a child's entire bedroom could be decorated in Hoppy. Perhaps a few actually were. Most parents drew the line at a few accessories mixed with generic items to create THE LOOK. One suspects parents knew that the craze would not last forever and that making a substantial investment, only to do it again in three or four years, made little sense. Of course, this logic stank in the mind of a ten year old child.

Nowhere is the quality of a Hoppy licensed product more evident than in those licensed for the bedroom, especially textiles. Boyd knew his products would be used and used hard. Nobody bought two - one to put away and another to play with - in the Fifties. Collecting was a hobby. Accepting the fact that textiles such as throw rugs and bedspreads were washed dozens, even hundreds of times, most survivors still show only minor wear, with colors remaining bright and strong.

Few collectors think about Boyd's marketing strategy. It had a clear focus - children between the ages of six and twelve. Boyd licensed little merchandise for infants and pre-schoolers. He recognized that by the time a child reached junior high school, he or she would be pulled in a myriad of directions, few of which included Hoppy.

Boyd was a shaper and molder of children. He wanted contact with them when his influence could make a difference.

Nowhere is Boyd's marketing strategy better reflected than in the furniture and furniture accessories that he licensed. Everything was designed in size for the elementary school cowboy and cowgirl. When a youngster outgrew it, it could be passed along to someone smaller. It was designed to last.

Hoppy's Bunk House. The following description includes only those items that are not pictured separately elsewhere in this book:

1. Bedroom suite, Morgan Manufacturing Company, Ashville, North Carolina. This ten piece Hopalong Cassidy bedroom suite consists of two beds that can be used separately or as bunk beds (the set came with four aluminum support rods when used as bunk beds and eight end post knobs), the "HOPALONG CASSIDY RANCH" sign for use on one of the headboards, ladder for the bed, a chest of drawers, framed mirror, fall front desk, desk chair, end table, and toy chest. Hoppy identification is minimal. The chest of drawers and desk have a piece of leather attached with a bust portrait of Boyd and Topper. The only other Hoppy tie-in is the sign for the bed headboard.
2. Curtains, Chenille, Tan, Bell Textile Company, Inc., New York, New York. Bell held the license for "Children's chenille bedspreads, bedroom rugs, draperies, bath mats and lids." Bell's curtains were designed for use as draw drapes. They were sold in pairs with distinct left and right units.
3/4. Bedspreads, Chenille, Tan and Yellow grounds, Bell Textile Company. The Hoppy bedspread is found in three basic background colors - tan, yellow, and white. All chenille products were tied together through color and the use of common decorative motifs.
5/6. Pillow Covers, Silk, Casey Premium Merchandise Company. Two silk pillow cover decorative motifs have been identified: (1) a bust portrait of Hoppy and (2) Hoppy astride Topper. A find of approximately a dozen excellent examples of each entered the market in the early 1980s.

7. Bath Mat, Chenille, White, Bell Textile Company. The bath mat is slightly larger and fluffier than the throw rug. Not included in the picture is a medium size chenille rug. This has a white ground and is often confused with a bedspread. The rug has a slightly different decorative motif and is much heavier.
8. Throw Rug, Tan, Bell Textile Company. The throw rug and bath mat shared the same decorative motif — a diagonal fence in the middle of which is a portrait of Topper's head and the words "HOPALONG CASSIDY." Tan is the most common background color for the throw rug. Other background colors do exist.
9. Clothes Hamper, Loroman Company, New York, New York. The Loroman license called for "metal or fibre" clothes hampers. It is questionable if the company ever marketed a fiber model.
10. Round-Up Clothes Corral or Clothes Rack, Dairy Premium. [Partially visible on the left just above the top of the desk.] This item apparently was one of several produced exclusively for use in dairy promotions. The logo or name of the dairy is found on the lower right. The instructions for its use are simple "Hang 'em Here, Cowpokes."
11. Floor Covering, linoleum, generic Western theme. Although this pictures shows a generic Western floor covering, a Hopalong Cassidy theme linoleum floor covering was licensed by Boyd. The Paulsboro Manufacturing Company, Philadelphia, Pennsylvania, held a license for "linoleum floor coverings and wainscoting." **Reproduction alert**. Unconfirmed rumors are making the rounds that a few reproductions have been made. Optimists talk of a warehouse find.

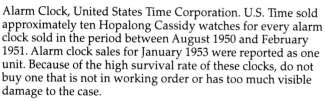

Alarm Clock, United States Time Corporation. U.S. Time sold approximately ten Hopalong Cassidy watches for every alarm clock sold in the period between August 1950 and February 1951. Alarm clock sales for January 1953 were reported as one unit. Because of the high survival rate of these clocks, do not buy one that is not in working order or has too much visible damage to the case.

Rocking Chair, manufacturer unknown. The rocker original came with a flat, wooden, silhouette horse's head (Topper) on a metal bracket that could be attached under the seat. A young child's legs straddle the head. As the child grew, the head usually was removed for convenience. The seat also came in blue vinyl.

T-V Chair, Transogram, New York, New York. Parents appreciated the portability of this folding T-V chair. Identical chairs can be found with other fabric themes such as a generic cowboy motif or with the image of another television Western cowboy hero.

Lamp, Rotating, Econolite Corporation, Los Angeles, California. Several versions of this lamp are know, both in respect to the rotating section as well as labeling on the plastic shade. Check carefully for scratching of the motif on the plastic shade and base deterioration of the rotating section before buying.

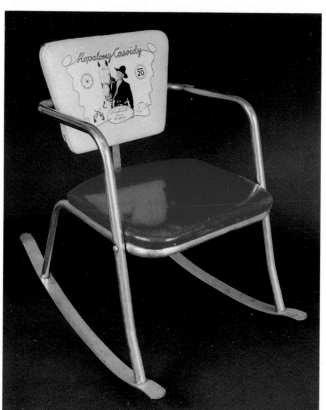

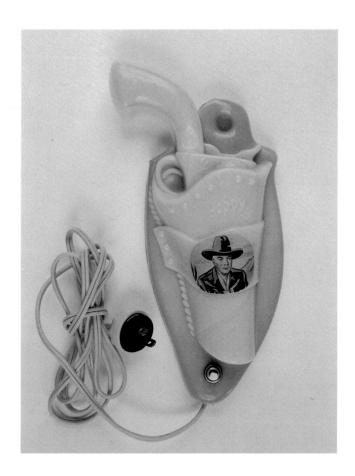

Lamp, Table, Aladdin Industries. This lamp is an excellent example where cross over collectors have driven up the price. Aladdin collectors are as fanatical, if not more so, than Hoppy collectors. This is a "must have" lamp in any Aladdin collection. What makes this lamp a Hoppy product? The answer is the shade which is unfortunately missing from this example.

Lamp, Wall, Aladdin Industries. This is another example where a Hopalong Cassidy decal was all that was required to turn this generic cowboy lamp into a Hopalong Cassidy product.

Pennant, Casey Premium Merchandise Company, Chicago, Illinois. This Hoppy felt pennant came in a large and small size. A large number survive in fine or better condition. Avoid buying damaged examples. **Reproduction alert**.

Picture Frame, manufacturer unknown. The sales insert was designed to allow the imprinting of the name of a local photographer. A Hoppy image and/or endorsement was used to sell a number of different types of picture frames.

Radio, Arvin Industries. The radio came in two versions - a red case and a black case. The die used to produce the foil covering on the radio is in private hands. The owner is making modern restrikes and selling them. When buying a radio, make certain that it includes its antenna.

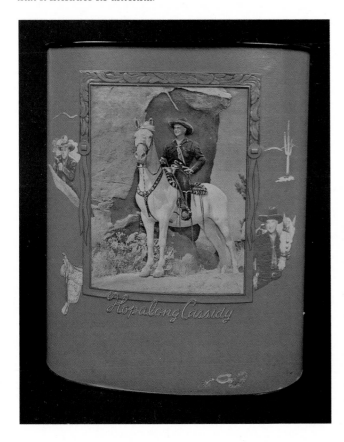

Towel, attributed to R. A. Briggs & Company, Chicago. Boyd issued two licenses for towels, one to R.A. Briggs for "children's towels" and a second to Star Embroidery Company of Los Angeles California for "wash cloths and towels." Surviving towels lack manufacturer labels making exact identification impossible.

Wastebasket, Hassenfeld Brothers, Pawtucket, Rhode Island. Hassenfeld Brothers used the same covering that appeared on their scrap books and photo albums and applied two of them to a metal container to make this wastebasket.

Fabric, manufacturer unknown. One of many fabrics with a Hopalong Cassidy theme available for purchase by the home sewer. Joseph Caro's *Collector's Guide to Hopalong Cassidy Memorabilia* pictures a pillow made from this fabric.

Tie Rack, Anson, Inc., Providence, Rhode Island. Anson received a license to make tie-holders along with its license to make boys' jewelry. This molded tie rack shows Anson's commitment to producing quality products. It was marketed under Anson's Slick Company tradename.

Wallpaper Roll, Harmony House. Only a few complete rolls of wallpaper are know. Once again, unscrupulous dealers who have acquired complete rolls have destroyed them by selling two to three feet sections to collectors. Hoppy wallpaper is found in several different designs and background colors.

Bedspread/Blanket, Belcraft Company, possibly a division of Bell Textiles. Bell's licensing agreement read: "You shall have, likewise, the rights pursuant to this agreement, to use any other materials which you consider appropriate in the manufacturer, sale and distribution of the aforementioned articles." This bedspread is made of light wool, not chenille.

Sing Along with Hoppy

I initially wanted to call my Hoppy book "The Good Guy Wore Black." Alas, he did not. Hoppy's shirt was a dark Navy blue and his striped pants also were dark blue. My next choice was "He Only Kissed His Horse." Unfortunately, Hoppy did kiss a woman in one of his movies. Moving on in the wonderful world of title selection, I thought about "He Never Sang A Note." While Hoppy did not sing, several of his romantic leads and others did. Further, music played an important role in the Hopalong Cassidy movies that such a title would be misleading.

William Boyd's image as Hopalong Cassidy was used to promote sheet music and music folios. Not all the music was background music from Hoppy films. Some was written to take advantage of the craze for Hoppy anything. A number of old favorites gained new life because of a picture of Hoppy on the cover.

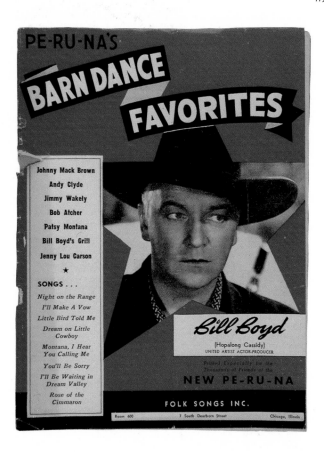

Song Folio, Hopalong Cassidy Musical Roundup, Nacio Publications Inc. and Consolidated Music Publishers, Inc. Among the titles are "I'm Saving Up To Buy A Saddle" and "'Hoppy Wishes You A 'Hoppy' Birthday." Obviously, these are songs written specifically for sale to the television Hoppy audience.

Song Folio, Pe-Ru-Na's Barn Dance Favorites, Folk Songs, Inc. The only thing Hoppy about this album is his picture on the cover.

Magazine Tear Sheet, origin unknown. Paramount distributed their Hopalong Cassidy film music folio through music dealers and via direct advertisement.

Song Folio, Hopalong Cassidy Song Folio, Paramount Music Corporation. This folio did feature songs and background music from Sherman's Hopalong Cassidy films.

Contemporary Licensed Products

U. S. Television Office, Inc., owns the character, literary, motion picture, radio, and all additional rights to the character Hopalong Cassidy. The organization continues to license Hopalong Cassidy products. Individuals and companies wishing to obtain information about licensing can write: U.S. Television Office, Inc., PO Box 387, Greenville, NY 12083.

Most of the items listed below are in current production and can be obtained through normal retail channels.

Collectors' Card Set, Riders of the Silver Screen, Series 1, Smoky Mountain

Belt Buckles, Smoky Mountain. Two versions of this belt buckle have been manufactured — one plain and the other with enamel decoration. Each was made in an edition of 10,000.

Hopalong Cassidy Colt 45, hand engraved. Premier Edition, Handgun, U.S. Historical Society. *Courtesy of the U.S. Historical Society.*

Hopalong Cassidy Single Action Army .45 Revolver, U.S. Historical Society. *Courtesy of the U.S. Historical Society.*

Key Chain, Bullet, Arvo Jola Products, Inc.

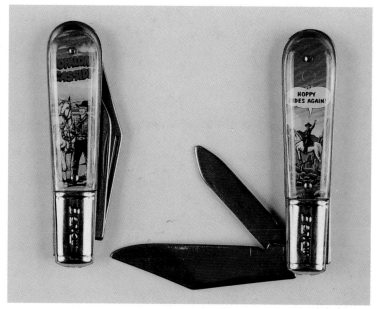

Pocket Knife, Smoky Mountain

Pocket Knife, Smoky Mountain

Limited Edition Collectors Plate, The Hamilton Collection

Hopalong Cassidy Pocket Knife from Riders of the Silver
Screen collection, Smoky Mountain

Sign, Arvin, Hopalong Cassidy Radio, Smoky Mountain

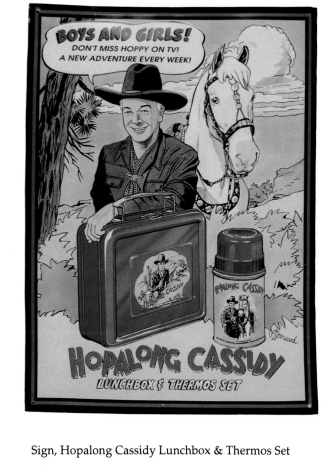

Sign, Hopalong Cassidy Lunchbox & Thermos Set

Sign, Bond Bread, Smoky Mountain

Record Album, Hopalong Cassidy Official Radio Broadcasts, George Garabedian.

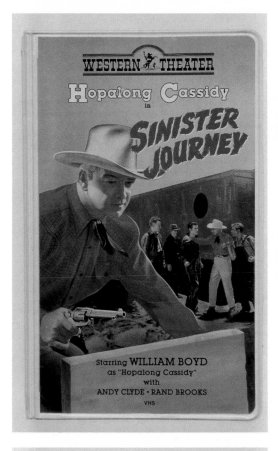

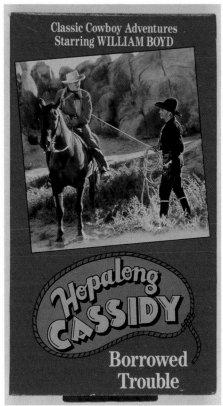

Video, Movie, Walt Disney, Buena Vista Home Video, present packaging, $9.99 retail.

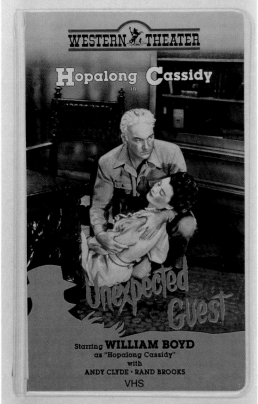

Video, Movie, Walt Disney, Buena Vista Home Video, early packaging, retail above $20.00.

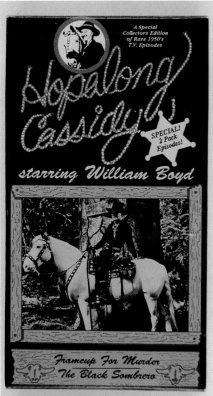

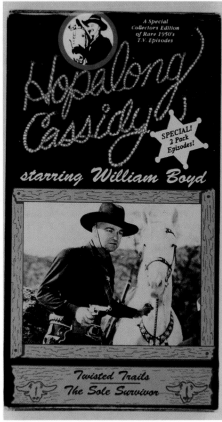

Video, Half Hour Television Programs, World Wide Video, Inc., distributed by Handleman Co.

Hoppy Festivals

Hopalong Cassidy was the featured cowboy hero at the Western Film Fair held July 6th through 8th, 1989, in Raleigh, North Carolina. U.S. Television Office, Inc., co-operated with festival promoters to permit the showing of several uncut Hopalong Cassidy feature movies.

In 1991 the Lone Pine Sierra Film Festival honored Hopalong Cassidy. A report about the festival and photographs of the event are found in Joseph Caro's *Collector's Guide to Hopalong Cassidy Memorabilia*.

1991 also marked the first year of the Hopalong Cassidy Festival in Cambridge, Ohio. The festival, held in late April or early May, has become an annual event. The festival enjoys the strong support of U.S. Television Office, Inc. For information about future festivals and their dates, write: Cambridge Chamber of Commerce, Cambridge, OH 43725.

The June 2-4, 1995 Lone Pine Sierra Film Festival honored the 100th anniversary of William Boyd's birth. Activities included visits to many of the locations where the Hoppy films were shot.

Pinback button, Lone Pine Sierra Film Festival, 1991

Pinback button, Western Film Fair, Raleigh, North Carolina, 1989

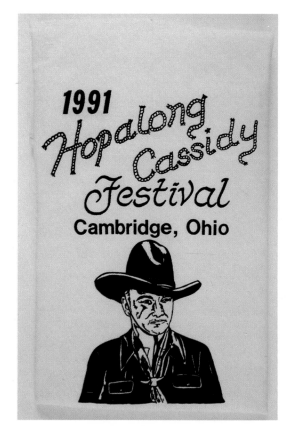

Logo, Hopalong Cassidy Festival, Cambridge, 1991

Logo, Hopalong Cassidy Festival, Cambridge, 1993

Logo Hopalong Cassidy Festival, Cambridge, 1992

T-shirt, Hopalong Cassidy Festival, Cambridge, 1991

Logo, Selection of items, Hopalong Cassidy Festival, Cambridge, 1994

T-shirt, Hopalong Cassidy Festival, Cambridge, 1994

Hat, Hopalong Cassidy Festival, Cambridge, date unknown

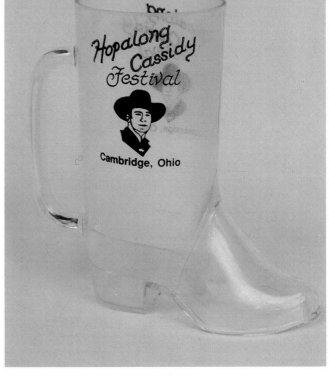

Drinking Glass, Boot, Hopalong Cassidy Festival, Cambridge, date unknown

Coffee Mugs, Hopalong Cassidy Festival, Cambridge. Left - date unknown; Right - 1991.

Key Ring, Hopalong Cassidy Festival, Cambridge, date unknown, possibly 1991

Key Ring, Hopalong Cassidy Festival, Cambridge, 1993

Paperweight, Hopalong Cassidy Festival, Cambridge, 1992

Toothpick, Boyd Crystal Art Glass, Cambridge, Ohio. Boyd is a major manufacturer of glass novelties. Typically the company offers a mold in a large variety of colors.

Stolen Treasures

Within the past twenty years and especially recently, the collectors' market has been plagued with unlicensed Hoppy reproductions, fantasy items, pirates, and fakes. The problem is epidemic. Numerous unsuspecting buyers, experienced collectors, and novices alike, have been stung.

The key is that these products are unlicensed. The owners of the rights to the Hopalong Cassidy character receive no royalty payment for the use of Hoppy's image. Instead, they must involve themselves in expensive lawsuits to make individuals involved with these products cease manufacture and distribution. In almost every instance, any award of money damages is thwarted by the offending individual or firm declaring bankruptcy.

You can help put these crooks out of business by (1) refusing to buy from them and (2) spreading the word about them and asking others to do likewise. If you must buy an example, do not pay more than a few dollars. These items are sold in quantity through reproduction houses for pennies. Their long term collectibility must be seriously questioned.

Recently, I received a call asking my opinion about the period authenticity of a Hopalong Cassidy pedal car. It never existed. The car was scheduled to be auctioned a few days later. This is an excellent example of the faking currently being done of Hopalong Cassidy material. The wagon illustrated below is another. These one-of-a-kind fakes are designed to appeal to the "high end" collector looking for the rarest of the rare. BEWARE!

What about one-of-a-kind items relating to Hoppy that someone has made or has had made as a present? These exist. The key is they were made and given with no intent to deceive. Such "novelties" enhance a collection. However, collectors do not go out of their way to obtain them nor pay large sums of money for them.

Especially bad are the video pirates. Their products are generally fair to awful in quality. Further, many of the films have sections missing, the videos having been pirated from the cut down television movies. Now that U.S. Television Office, Inc., is involved in a program to provide full length releases of all Hopalong Cassidy movies and television shows based on original masters, buying pirate videos makes absolutely no sense.

Badge, fantasy piece. The sheriff's badge is a totally new product. It is not a copy of a period example.

Wagon, fake. Boyd never licensed a wagon. A period wagon was repainted in a classic 1950s Hoppy motif. Note the use of a period Hoppy conch from a holster as a means of adding a degree of authenticity to this fake.

Reply Card, reproduction. This card along with many of the paper items associated with the Hopalong Cassidy Savings Club have been reprinted. They are extremely difficult to tell from the period pieces.

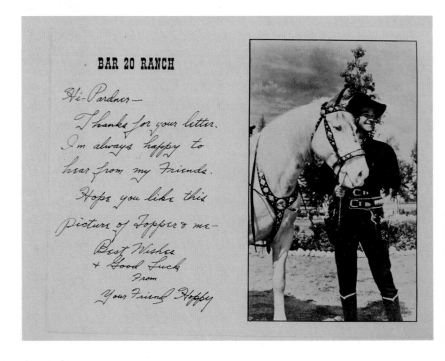

Good Luck Token, period and reproduction. In this instance, the strongly cast object on the left is the reproduction. The period token, center and right, is light in weight. Note the differences in the two facial treatments.

Lid, plastic, fantasy item. The inspiration for this fantasy item was a period product lid.

Sheriff's Badges, period and reproduction. The period badges are on the right and in the middle. Note the sharp detail of the casting, the lack of impression of the front image on the back, and the manner by which the pin is attached. The reproduction of the period badge is on the right. If you cannot tell the difference between the two from the front, check the back.

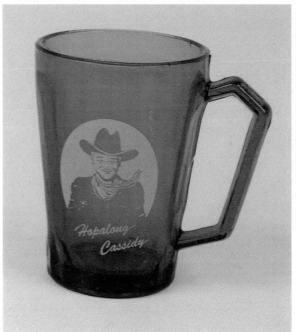

Child's Glass Mug, fantasy item. No Hoppy did not have thing going with Shirley Temple. This mug mimics a 1930s Shirley Temple premium mug.

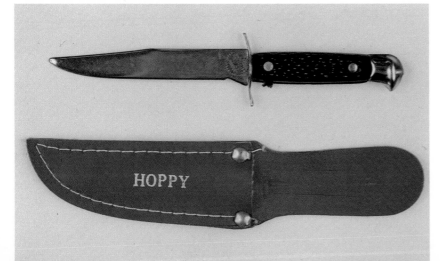

Knife and Sheath, fantasy item. There is little to stop anyone from simply printing Hoppy' name on something.

Bottle Cap, plastic, and Neckerchief Slide, metal, fantasy items. These are brand new products. Nothing like them existed in the 1950s.

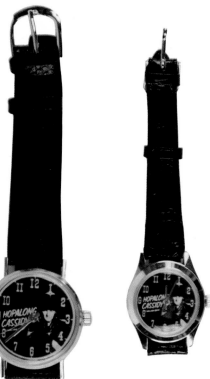

Good Luck Horseshoe, fantasy item. This is a complete fabrication I have seen it offered by dealers in the field at prices beginning at $25.00 and ranging up to $45.00. One can buy it wholesales for pennies.

Wrist Watches, fantasy items. These are cheap foreign imports - cheap in manufacture and highly unreliable.

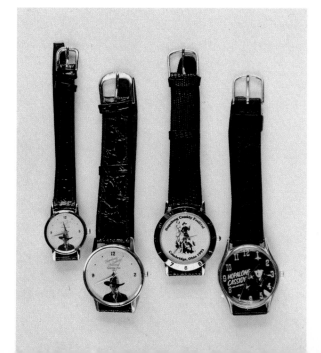

Wrist Watches, fantasy items and licensed product. U.S. Television Office, Inc., did license the third watch from the left as a souvenir item to be sold at the 1992 Hopalong Cassidy Festival in Cambridge, Ohio. The other watches are unlicensed.

Marbles, fantasy item. These marbles have fooled some of the leading marble collectors in the country. They are not period.

Pocket Watch, fantasy item. U.S. Time did produce a Hopalong Cassidy pocket watch in the early 1950s. It utilized U.S. Time's Hopalong Cassidy wrist watch motif. This watch is designed to fool the unknowledgeable buyer.

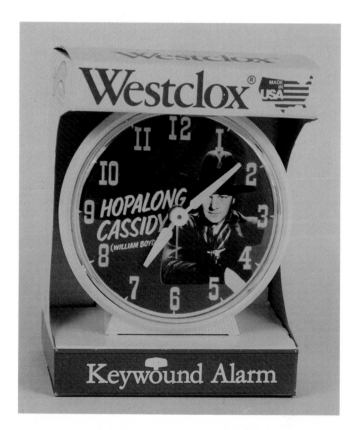

Alarm Clock, fantasy item. This is a knockoff. The packaging contains a name that sounds similar to a major clock manufacturer but is not.

Video, Movie, pirate. Cumberland Video is one of the biggest pirates of Hopalong Cassidy videos.

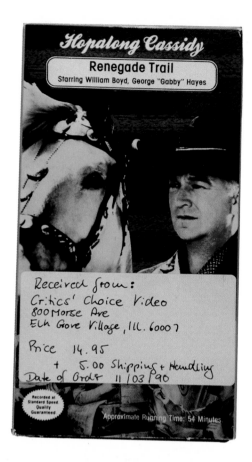

Video, Movie, pirate. Two videos obtained from Critics' Choice Video. The documentation on the one is for potential use in a law suit. The videos were wholesaled to Critics' Choice Video by Hollywood Select, another major Hopalong Cassidy video pirate.

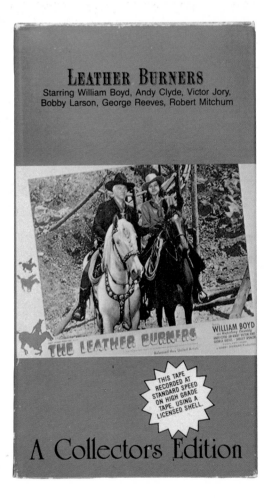

Video, Movie, pirate. Many pirates simply forget to have their name printed on the video jackets. By the time their products are spotted, they are in the hands of distributors (a highly uncooperative group) or on store shelves. This makes the task of finding them increasingly difficult.

Not Quite Hoppy or Not Quite Certain

While Boyd could license the Hoppy image and name, he could not control the use of the image of a generic cowboy dressed in black and riding a white horse. As a result, there were a number of period Hopalong Cassidy copycats. While different enough to avoid infringement questions, these products probably confused their fair share of parents and kids.

Even with the wealth of documentation that exists about Boyd's product licensing, one is not certain about some items. Confusion rests primarily with the products produced by premium manufacturers such as Einson-Freeman. Unfortunately, Boyd did not retain an example of every product that the licensed. As a result, there are a number of items that fall into the "your guess is as good as mine" category.

Wastebasket, manufacturer unknown, copycat. How close to Hoppy can you get?

Gloves, manufacturer unknown, copycat. The cowboy image is enclosed in a lariat frame, a motif that was used continually in Hoppy images.

Neckerchief, manufacturer unknown, copycat. Extremely close in respect to material and motif to a Hopalong Cassidy white rayon neckerchief/scarf.

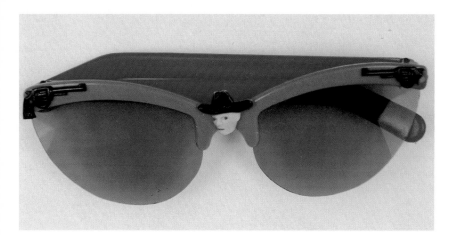

Sunglasses, manufacturer unknown. I vote not period; what's your guess?

Charm, manufacturer unknown. They look too good and too new to be old. However???

Badge and Sheriff's Badge, manufacturer unknown. I had to include this picture because only someone from abroad would fail to realize that the letters for HOPALONG on the compass badge have been reversed. Really no question in my mind about these two items.

HOPALONG CASSIDY LICENSEES

Included among the William Boyd papers are more than half a dozen lists of licensees for Hopalong Cassidy merchandise. The following listing combines these lists into one master list.

There are only two limiting factors to this listing. First, it contains only licenses granted by William Boyd. The Mulford estate and Doubleday owned the literary rights to Hopalong Cassidy and obviously granted a number of such licenses during the Hoppymania of the late 1940s and early 1950s. The Boyd papers did not contain a list of these licenses. Second, the earliest list dates from 1950, not 1948/49. Several licensees changed prior to the 1950 date. As a result, a few early licensees may be missing from this master listing.

Each list is dated. The date of the list(s) on which a licensee appears is noted in the heading: "List of Licensees."

This is not the date nor the term of the license. When known this information appears in the "Date of Contract" heading.

In several instances you will note multiple listings for a licensee, usually the result of a company moving or using its show room address one time and plant address another. All references are given to make the list as accurate as possible.

Not every licensee appears on every list. Most licenses had expired years earlier when Marguerite Cherry prepared the October 19, 1962 list, a master record of sorts for William Boyd.

If you have additional information about any of these licensees and wish to share it, send it to: Harry L. Rinker, 5093 Vera Cruz Road, Emmaus, PA 18049. A great big Hoppy thanks!

AMERICAN LICENSEES

FOOD PRODUCTS

BAKERIES: A general contract with Quality Bakers of America Cooperative, Inc., 120 West 42nd Street, New York, NY, was signed in 1951. Member bakeries are noted with *QBA* following the date of contract.

Aikman Bread Company
1301 Eleventh Street
Port Huron, MI
Territory: MI counties of St. Clair, Sanilac, Huron, Lapeer, Tuscola
Date of Contract: 7/26/52 QBA
List of Licensees: 10/19/62

Atlanta Baking Company
165 Bailey Street S.W.
Atlanta, GA
Territory: GA area
Date of Contract: 1/10/52
List of Licensees: 11/15/51, 7/10/52, 10/19/62

Bamby Bakers (two plants)
Salisbury, NC
Burlington, NC
Territory: NC counties of Howan, Wake, Chatham, Moore, Davie, Lee, Union, Davidson, Stanley, Cumberland, Guilford, Cabarrus, Randolph, Hoke, Robeson, Forsyth/Rockingham, Scotland, Harnett, Alamance, Richmond, Durham, Orange, Anson
Date of Contract: 7/18/52 QBA
List of Licensees: 10/19/62

Bohnet Baking Company
219 Meerscheidt Street
San Antonio 6, TX
Territory: TX area
Licensed: Mrs. Bohnet's Bread
Date of Contract: 12/18/51
List of Licensees: 1/3/52, 7/10/52, 10/19/62

Boge Bros. Bakery
S. 401 Sherman Street
Spokane 10, WA
Territory: WA city of Spokane, ID city of Coeur d'Alene
Date of Contract: 8/3/54
List of Licensees: 2/8/55, 3/1/55, 1/20/56, 6/20/56, 10/19/62

Bowman's Bakery
102 Park Street, NW
Roanoke, VA
Territory: VA counties of Roanoke, Botetourt, Bedford, Pittsylvania, Franklin, Henry, Patrick, Carroll, Pulaski, Montgomery, Giles, Craig
Date of Contract: 6/11/52 QBA
List of Licensees: 7/10/52, 10/19/62

Burry Biscuit Corporation
925 Newark Avenue
Elizabeth 3, NJ
Territory: Entire United States
Licensed: Cookies including butter flavored and chocolate crisp
Date of Contract: 7/1/50
List of Licensees: 11/15/51, 1/3/52, 7/10/52, 10/19/62

Butter Krust Bakeries, Inc.
1102 South Florida Avenue
Lakeland, FL
Territory: FL counties of Polk, Hernando, Pasco, Hardee, Highlands, Manatee, DeStoto, Sarasota, Orange, Charlotte, Osceola, Sumter, Marion, Hillsboro, Pinellas
Date of Contract: 6/15/52 QBA
List of Licensees: 7/10/52, 10/19/62

Campbell-Sell Baking Company
1125 Twelfth Street
Denver, CO
Territory: CO counties of Adams, Arapahoe, Boulder, Custer, Douglas, Eagle, Elbert, El Paso, Fremont, Gilpin, Grand, Jackson, Jefferson, Larimer, Logan, Morgan, Park, Phillips, Pueblo, Sedgwick, Summit, Teller, Weld; NB counties of Banner, Cheyenne, Devel, Garden, Keith, Kimball, Morrill, Scotts Bluff; WY counties of Albany, Goshen, Laramie, Platte
Date of Contract: 10/29/52 QBA
List of Licensees: 11/15/51, 10/19/62

Firsch Baking Company, Inc.
1902 Cranberry Street

Erie, PA
Territory: OH county of Ashtabula; PA counties of Crawford, Erie
Date of Contract: 6/2/52 QBA
List of Licensees: 7/10/52, 10/19/62

Fisher Baking Company
1844 S. State Street
Salt Lake City 10, UT
Territory: Entire state of UT
Date of Contract: 10/15/51
List of Licensees: 11/15/51, 1/3/52, 7/10/52, 10/19/62

Flowers Baking Company, Inc.
Madison Street
Thomasville, GA
Territory: GA counties of Alabama, Florida
Date of Contract: 7/15/52 QBA
List of Licensees: 10/19/62

General Baking Company
420 Lexington Avenue
New York 17, NY
Territory: Entire state of NY
Licensed: Bond Bread
Date of Contract: 5/21/51
List of Licensees: 11/15/51, 1/3/52, 7/10/52, 10/19/62

Hardin's Bakery, Inc.
Tuscaloosa, AL
Territory: AL counties of Franklin, Winston, Walker, Fayette, Lamar, Pickens, Bibb, Tuscaloosa, Hale, Jefferson, Green, Marengo, Perry, Shelby, Marion, Lauderdale, Colbe
Date of Contract: 6/26/52 QBA
List of Licensees: 10/19/62

Hecht's Bakery, Inc.
Seventh & Shelby Streets
Bristol, TN
Territory: VA counties of Wythe, Smythe, Washington, Russell, Scott, Dickinson, Grayson, Wise, Lee; TN counties of Sullivan, Johnson, Washington, Unicoi; KY counties of Letcher, Harlan
Date of Contract: 8/30/52 QBA
List of Licensees: 10/19/62

Holsum Baking Company
220 Tennessee Street
Pine Bluff, AK
Territory: 27 counties in state of AK
Date of Contract: 6/24/52 QBA
List of Licensees: 10/19/62

Langendorf United Bakeries, Inc.
1160 McAllister Street
San Francisco, CA
Territory: Entire states of CA, OR, and WA; NE county of Washoe
Licensed: Barbara Ann Bread
Date of Contract: 1/1/52
List of Licensees: 11/15/51, 1/3/52, 7/10/52, 10/19/62

Liberty Baking Company
6018 Houston Street
Pittsburgh, PA
Territory: VA counties of Beaver, Butler, Armstrong, Westmoreland, Fayette, portion of Allegheny
Date of Contract: 7/18/52 QBA
List of Licensees: 10/19/62

Love's Biscuit & Bread Company
P.O. Box 294
Honolulu, HI
Territory: All Hawaiian Islands
Date of Contract: 5/15/52
List of Licensees: 7/10/52, 2/8/55, 3/1/55, 10/19/62

Lowenberg Bakery, Inc.
Ottumwa, IA
Territory: IA counties of Illinois, Missouri
Date of Contract: 8/5/52 QBA
List of Licensees: 10/19/62

Oak Cliff Baking Company
546 East 9th Street
Dallas, TX
Territory: TX cities of Dallas, Fort Worth
Licensed: Holsum Bread
Date of Contract: 5/29/51
List of Licensees: 11/15/51, 1/3/52, 7/10/52, 10/19/62

Oswald Jaeger Baking Company
918 West Somers Street
Milwaukee, WI
Territory: WI counties of Dane, Dodge, Columbia, Fond duLac, Green, Jefferson, Kenosha, Milwaukee, Ozaukee, Racine, Rock, Sheboygan, Walworth, Washington, Waukesha
Date of Contract: 3/28/52 QBA
List of Licensees: 7/10/52, 10/19/62

Pan-O-Gold Baking Company
Pipestone, MN
Territory: Various counties of IA, SD, MN
Date of Contract: 8/1/52
List of Licensees: 10/19/62

Peerless Bread Company, Inc.
837 North Main Street
Jacksonville, IL
Territory: IL counties of Morgan, Scott, Macoupin, Pike, Brown, Sangamon, Schuyler, Greene
Date of Contract: 7/23/52 QBA
List of Licensees: 10/19/62

Peerless Bakery
584 Higuera Street

Jacksonville, IL
Territory: CA county of San Luis Obispo
Date of Contract: 12/18/52
List of Licensees: 10/19/62

Quality Bakers of America Cooperative, Inc.
120 W. 42nd Street
New York, NY
Date of Contract: 1951
List of Licensees: 11/15/51, 7/10/52

Rushmore Baking Company
Rapid City, SD
Territory: SD counties of Fall River, Custer, Lawrence, Meade, Pennington
Date of Contract: 6/20/55
List of Licensees: 1/20/56, 6/20/56, 10/19/62

Sabine Holsum Bakers
713 Houston Avenue
Port Arthur, TX
Territory: TX area in and about Beaumont and Port Arthur
Date of Contract: 3/2/52
List of Licensees: 11/15/51, 7/10/52, 10/19/62

Sam Schneider Bakery
302 High Street
Memphis, TN
Territory: TN counties of Shelby and Crittenden and city of Memphis
Date of Contract: 2/20/52
List of Licensees: 11/15/51, 7/10/52, 10/19/62

Schmidt Baking Company, Inc.
800 Frederick Street
Cumberland, MD
Territory: PA counties of Armstrong, Bedford, Blair, Cambria, Fayette, Indiana, Mifflin, Somerset, Westmoreland; WV counties of Mineral, Preston; MD counties of Allegany, Garrett
Date of Contract: 5/29/52
List of Licensees: 7/10/52, 10/19/62

Sta-Kleen Bakery, Inc.
Danville, VA
Territory: VA counties of Patrick, Henry, Pittsylvania, Halifax, Charlotte, Mecklenburg
Date of Contract: 7/15/52 QBA
List of Licensees: 10/19/62

Sta-Kleen Bakery, Inc.
Lynchburg, VA
Territory: VA counties of Campbell, Appomattox, Charlotte, Prince Edward, Amherst, Bedford, Pittsylvania, Nattoway, Halifax, Allegheny
Date of Contract: 7/15/52 QBA
List of Licensees: 10/19/62

Stroehmann Brothers Company
Sayre, PA
Territory: NY cities of Elmira, Binghamton, and areas around Schuyler, Chemung, Tioga, Broome; PA city of Sayre and county of Bradford
Date of Contract: 4/15/52 QBA
List of Licensees: 10/19/62

Stroehmann Brothers Company
Williamsport, PA
Territory: 25 PA counties, 1 NY county
Date of Contract: 7/18/52 QBA
List of Licensees: 7/10/52, 10/19/62

Swan Brothers, Inc.
1801 E. Magnolia Avenue
Knoxville, TN
Territory: TN counties of Anderson, Bledsoe, Blount, Bradley, Carter, Campbell, Claiborne, Cocke, Cumberland, Grainger, Greene, Hamblen, Hancock, Hawkins,

Jefferson, Knox, Loudon, McMinn, Meigs, Monroe, Morgan, Polk, Roane, Scott, Sevier, Union
Date of Contract: 6/30/52 QBA
List of Licensees: 10/19/62

Waldensian Bakers, Inc.
Valdese, NC
Territory: NC counties of Alexander, Alleghany, Ashe, Avery, Buncomb, Burke, Caldwell, Catawaba, Graham, Haywood, Henderson, Iredel, Jackson, Lincoln, Macon, Madison, Mitchell, McDowell, Rutherford, Polk, Surray, Swain, Transylvania, Watauga, Wilkes, Yadkin, Yancy, Cleveland, Gaston, Mecklenburg; SC counties of Cherokee, Spartanburg, Greenville, Pickens, York, Union, Anderson, Laurens, Newberry
Date of Contract: 7/21/52 QBA
List of Licensees: 10/19/62

Waldensian-Bitt Baking Company, Inc.
1602 Dickenson Avenue
Greenville, NC
Bakery sold to **Columbia Bakers**
Territory: NC counties of Pitt, Beaufort, Pamlico, Lenoir, Wayne, Greene, Wilson, Edgecomb, Martin, Wash, Craven, Washington, Hyde
Date of Contract: 7/21/52 QBA
List of Licensees: 10/19/62

DAIRIES: An asterisk (*) denotes those dairies which also had an ice cream franchise.

All Star Dairy Association
Suite 1205
295 Madison Avenue
New York, NY
Licensed: Milk
List of Licensees: 12/1/58

American Dairy Company, Inc. *
700 East Missouri Street
Evansville, IN
Territory: IN counties of Dubois, Gibson, Posey, Perry, Spencer, Vanderburg, Warrick; IL counties of Edwards, Wabash, White; KY counties of Davies, Hancock, Henderson, Union
Licensed: Milk, ice cream, cottage cheese
Date of Contract: 1/15/55
Termination Date: 2/28/58
List of Licensees: 3/15/55, 1/20/56, 6/20/56, 10/1/56, 2/25/57, 9/24/56, 2/14/58, 10/19/62

American Dairies, Inc. *
P.O. Box 286
Kansas City 41, MO
Territory: KS counties of Jackson, Shawnee, Wabaunsee, Osage, Jefferson, Pottawatomie
Licensed: Milk, ice cream
Date of Contract: 6/10/54
List of Licensees: 10/19/62

Anderson Dairy *
824 South Fifth Street
Las Vegas, NE
Territory: NE counties of Clark, Lincoln, Nye, Esmeraldo, Mineral
Licensed: Milk, ice cream
Date of Contract: 6/15/55
List of Licensees: 1/20/56, 6/20/56, 10/19/62

Bowman Farm Dairy *
Fish Hatchery Road
Madison 5, WI
Territory: WI county of Dane
Licensed: Milk, ice cream

Date of Contract: 6/2/59
List of Licensees: 10/19/62

Brown Ice Cream & Milk Company *
823 Adams Street
Bowling Green, KY
Territory: KY counties of Allen, Barren, Butler, Edmonson, Logan, Muhlenberg, Ohio, Simpson, Todd, Warren
Licensed: Milk, ice cream
Date of Contract: 3/20/55
Termination Date: 6/30/58
List of Licensees: 1/20/56, 6/20/56, 10/1/56, 2/25/57, 9/24/57 2/14/58, 10/19/62

Brown's Velvet Ice Cream Corp. *
1332 Baronne Street
New Orleans, LA
Territory: All LA parishes east and south of and including E. Feliciana, E. Baton Rouge, W. Baton Rouge, Iverville, St. Martin, Lafayette, Stone, Hancock, Harrison, Pearl River
Licensed: Milk, ice cream
Date of Contract: 6/18/54
Termination Date: 8/31/57
List of Licensees: 2/8/55, 3/15/55, 1/20/56, 6/20/56, 10/1/56, 2/25/57, 10/19/62

Butler's Dairy
Box 288
Madison, IN
Territory: IN counties of Jefferson, Switzerland; KY counties of Trimble, Carroll
Licensed: Milk, cottage cheese
Date of Contract: 8/30/54
Termination Date: 10/31/57
List of Licensees: 2/8/55, 3/15/55, 1/20/56, 6/20/56, 10/1/56, 2/25/57, 10/19/62

Cabarrus Creamery Company, Inc. *
357 North Church Street
Concord, NC
Territory: NC counties of Cabarrus, Rowan, Stanly
Licensed: Milk, ice cream, cottage cheese
Date of Contract: 1/15/55
Termination Date: 3/31/58
List of Licensees: 3/15/55, 1/20/56, 6/20/56, 10/1/56, 2/25/57, 9/24/56, 2/14/58, 10/19/62

Carolina Dairy, Inc.*
Grover & Edgemont Streets
Shelby, NC
Territory: NC counties of Cleveland, Polk, Rutherford; SC counties of Cherokee, Spartanburg
Licensed: Milk, ice cream, cottage cheese
Date of Contract: 2/7/55
Termination Date: 4/30/58
List of Licensees: 3/15/55, 1/20/56, 6/20/56, 10/1/56, 2/25/57, 10/19/62

Carolina Dairy Products Company, Inc.
Washington Street
Greenville, NC
Territory: NC counties of Pitt, Greene, Craven, Lenoir, Beaufort
Licensed: Milk
Date of Contract: 3/6/56
Termination Date: 4/30/59
List of Licensees: 6/20/56, 10/1/56, 2/25/57, 9/24/57, 2/14/58, 10/19/62

Cass-Clay Creamery, Inc.
Box 1246
Fargo, ND
Licensed: Milk
List of Licensees: 12/1/58

Cass-Clay Cooperative Creamery *
Moorhead, MN
Licensed: Milk, ice cream, cottage cheese
List of Licensees: 2/8/55, 3/15/55, 1/20/56, 6/20/56, 10/1/56, 2/25/57, 9/24/57

Chappell Creamery
118 North Fourth Street
Manhattan, KS
Territory: KS counties of Clay, Decatur, Marshall, Norton, Phillips, Riley, Sheridan, Thomas, Washington, Geary, Pottawatomie
Licensed: Milk, cottage cheese
Date of Contract: 6/11/54
List of Licensees: 2/8/55, 3/15/55, 1/20/56, 6/20/56, 10/1/56, 2/25/57, 9/24/57, 2/14/58, 10/19/62

Clarke Sanitary Dairy *
Gallup, NM
Territory: AZ county of Apache; NM counties of McKinley, San Juan; CO county of Montezuma
Licensed: Milk, ice cream
Date of Contract: 3/26/55
Termination Date: 6/30/58
List of Licensees: 1/20/56, 6/20/56, 10/1/56, 2/25/57, 9/24/57, 2/14/58, 10/19/62

Cloverlake Dairy Foods *
711 Austin Street
Plainview, TX
Territory: TX counties of Hale, Floyd, Lamb, Crosby, Castro, Cottle, Lubbock, Hockley, Motley, Swisher
Licensed: Milk, ice cream, cottage cheese
Date of Contract: 9/25/54
Termination Date: 11/30/57
List of Licensees: 2/8/55, 3/15/55, 1/20/56, 6/20/56, 10/1/56, 2/25/57, 9/24/57, 10/19/62

Cobble Dairy Products, Inc. *
Lexington, NC
Territory: Entire states of NC and SC, VA counties of Carroll, Floyd, Grayson, Halifax, Henry, Patrick, Pittsburg, Charlotte, Lunenburg, Mecklenburg, Brunswick
Licensed: Milk, ice cream
Date of Contract: 12/1/53
List of Licensees: 10/19/62

Country Club Dairy Company *
5633 Troost Avenue
Kansas City, MO
Territory: KS counties of Wyandotte, Miami, Johnson; MO counties of Johnson, Jackson, Clay, Platte, Cass, Lafayette, Ray
Licensed: Milk, ice cream, cottage cheese
Date of Contract: 12/14/54
Termination Date: 2/28/57
List of Licensees: 2/8/55, 3/15/55, 1/20/56, 6/20/56, 10/1/56, 10/19/62

Crombie's Ideal Dairy, Inc.
501 Second Avenue
Joliet, IL
Territory: IL county of Will
Licensed: Milk, cottage cheese
Date of Contract: 9/24/54
List of Licensees: 2/8/55, 3/15/55, 1/20/56, 6/20/56, 10/1/56, 2/25/57, 9/24/57, 2/14/58, 10/19/62

Dairyman's League Cooperative Association, Inc. *
100 Park Avenue
New York, NY
Territory: Entire state of NY; PA counties of Wayne, Lackawanna, Susquehanna, Wyoming, Luzerne, Schuylkill, North Cumberland, Columbia, Union, Montuer, Clinton, Lycoming, Sullivan, Tioga, Bradford, Carbon; NJ counties of Bergen, Passaic, Essex, Union, Hudson
Licensed: Milk, ice cream
Date of Contract: 4/1/55
Termination Date: 3/31/59, 4/31/59
List of Licensees: 1/20/56, 6/20/56, 10/1/56, 2/25/57, 9/24/57, 12/1/58, 10/19/62

Dairymens Ohio Farmers Milk Company *
3068 West 106th Street
Cleveland 11, OH
Territory: OH counties of Cuyahoga, Lorain, Lake, Medina
Licensed: Milk, ice cream
Date of Contract: 3/5/55
Termination Date: 5/31/58
List of Licensees: 1/20/56, 6/20/56, 10/1/56, 2/25/57, 10/19/62

Garden Farm Dairy, Inc. *
Box 239
East 60th & Albion Streets
Denver, CO
Territory: CO counties of Routt, Jackson, Larimer, Weld, Logan, Sedgwick, Phillips, Yuma, Morgan, Washington, Boulder, Grand, Eagle, Summit, Gilpin, Jefferson, Adams, Arapahoe, Pitkin, Lake, Park, Douglas, Elbert, Lincoln, Kiowa, Kit Carson, Cheyenne, Crowley, El Paso, Teller, Chaffee, Custer, Fremont, Pueblo, Otero, Bent, Prowers, Baca, Las Animas, Huerfano, Clerk Creek
Licensed: Milk, ice cream, cottage cheese
Date of Contract: 6/3/54
List of Licensees: 2/8/55, 3/15/55, 10/19/62

Gardiner Dairy & Ice Cream Company
Garden City, KS
Territory: KS counties of Finney, Ford, Grant, Scott, Seward, Stevens, Gray, Haskell, Meade, Kearny, Lane, Ness, Hodgeman
Licensed: Milk
Date of Contract: 7/31/54
List of Licensees: 2/8/55, 3/15/55, 1/20/56, 6/20/56, 10/1/56, 2/25/57, 9/24/57, 2/14/58, 10/19/62

Hannibal Quality Dairy, Inc. *
1248 Lyon Street
Hannibal, MO
Territory: MO counties of Audrain, Macon, Marion, Monroe, Pike, Ralls, Randolph, Shelby, Boone
Licensed: Milk, ice cream
Date of Contract: 5/1/54
Termination Date: 4/30/57
List of Licensees: 2/8/55, 3/15/55, 1/20/56, 6/20/56, 10/1/56, 2/25/57, 10/19/62

Harmony Farms, Inc.
2050 E. Main St.
Columbus, OH
Territory: OH counties of Franklin, Knox, Morrow, Licking, Richland, Marion, Delaware
Licensed: Milk, cottage cheese
Date of Contract: 12/20/54
Termination Date: 2/28/58
List of Licensees: 3/15/55, 1/20/56, 6/20/56, 10/1/56, 2/25/57, 9/24/57, 2/14/58, 10/19/62

Hawk Dairies, Inc. *
2429 East Eleventh Street
Tulsa, OK
Territory: Entire state of OK except counties of Harper, Ellis, Beckman, Roger Mills, Harmon, Jackson, Greet, Kiowa, Tillman, Cotton; KS counties of Chautauqua, Montgomery, Labette, Puritan
Licensed: Milk, ice cream, cottage cheese
Date of Contract: 6/4/54
List of Licensees: 2/8/55, 3/15/55, 10/19/62

Hawthorne-Mellody Farms
4224 W. Chicago Avenue
Chicago 51, IL
Territory: IL counties of Boone, Winnebago, McHenry, Lake, DeKalb, Kane, Cook, DuPage, Kendall, Kankakee; IN counties of Lake, Porter, LaPorte, St. Joseph, Elkhart; MI counties of Cass, Van Buren, Berrien; WI counties of Dodge, Columbia, Dane, Jefferson, Waukesha, Milwaukee, Greene, Rock, Walworth, Racine, Kenosha, Washington, Ozaukee
Licensed: Milk
Date of Contract: 5/10/55
List of Licensees: 1/20/56, 10/19/62

Hol-Guerns Dairy, Inc.
Canton, OH
Territory: OH city of Canton and the portion of Stark County within 7 mile radius of Canton
Licensed: Milk
Date of Contract: 3/7/52
List of Licensees: 7/10/52, 10/19/62

Home Milk & Ice Cream Company
c/o Marshall Ford Adv. Agcy.
918 Eleventh Street
Sacramento, CA
Territory: CA counties of Sacramento, Yolo, Placer, Yuba
Licensed: Milk
Date of Contract: 9/1/52
List of Licensees: 10/19/62

Hudson Dairy Company
129 N. Race Street
Urbana, IL
Territory: IL counties of Champaign, Vermilion
Licensed: Milk
Date of Contract: 8/38/58
List of Licensees: 10/19/62

Hyde Park Dairies, Inc *
943 S. McLean Boulevard
Wichita, KS
Territory: KS counties of Butler, Barber, Barton, Cowley, Harvey, Harper, Kingman, McPherson, Reno, Rice, Pratt, Sedwick, Sumner, Stafford
Licensed: Milk, ice cream, cottage cheese
Date of Contract: 3/4/53
List of Licensees: 2/8/55, 3/15/55, 1/20/56, 6/20/56, 10/1/56, 2/25/57, 9/24/57, 12/1/58, 10/19/62

Ideal Dairy Farms
2331 Morris Avenue
Union, NJ
Territory: NJ counties of Essex, Hudson, Middlesex, Morris, Passaic, Union
Licensed: Milk
Date of Contract: 2/11/55
List of Licensees: 3/15/55, 1/20/56, 10/19/62

Inland Empire Dairy Association
W. 1803 Third Avenue
Spokane 10, WA
Territory: WA city of Spokane
Licensed: Milk
Date of Contract: 12/1/51
List of Licensees: 11/15/51, 1/3/52, 7/10/52, 10/19/62

Jensen's Creamery *
Second & Topeka
Topeka, KS
Licensed: Milk, ice cream, cottage cheese
List of Licensees: 2/8/55, 3/15/55

Jersey Maid Corporation
Park Avenue
Bordentown, NJ
Territory: NJ counties of Mercer, Burlington, Cumberland, Camden
Licensed: Milk
Date of Contract: 2/2/55
List of Licensees: 1/20/56, 10/19/62

Jo-Mar Dairies Company *
1300 Iron Avenue
Salina, KS
Territory: KS counties of Morris, Dickinson, Marion, Salina, Ottawa, Cloud, Mitchell, Lincoln, Ellsworth, Osborne, Russell, Ellis, Rush, Pawnee, Edwards, Kiowa, Commanche, Clark, Geary
Licensed: Milk, ice cream, cottage cheese
Date of Contract: 12/3/54
List of Licensees: 2/8/55, 3/15/55, 1/20/56, 6/20/56, 10/1/56, 2/25/57, 9/24/57, 2/14/58, 10/19/62

Jorgensen Dairy Products *
P.O. Box 1467
Medford, OR
Licensed: Milk, ice cream
List of Licensees: 2/8/55, 3/15/55, 1/20/56, 6/20/56, 10/1/56, 9/24/57, 12/1/58

Klinke-Reed Dairy
1039 S. Bellevue
Memphis, TN
Territory: AK county of Crittenden; TN counties of Shelby, Madison, Chester, McNairy, Haywood, Crockett, Tipton, Lauderdale, Hardeman, Hardin, Henderson, Gibson, Fayette
Licensed: Milk, cottage cheese, orange drink
Date of Contract: 10/3/54
List of Licensees: 2/8/55, 3/15/55, 10/19/62

Kristoferson's Dairy
1300 Rainier Avenue
Seattle 44, WA
Territory: WA counties of King, Snohomish, and portion of Kitsap in Bainbridge Island
Licensed: Milk
Date of Contract: 9/15/51
List of Licensees: 11/15/51, 1/3/52, 7/10/52, 10/19/62

Lakeside Dairy Company
415 Broadway
Vallejo, CA
Territory: CA counties of Napa, Solano
Licensed: Milk, cottage cheese, orange drink
Date of Contract: 5/1/51
List of Licensees: 11/15/51, 1/3/52, 7/10/52, 3/15/55, 1/20/56, 6/20/56, 10/19/62

Lawrence Sanitary Milk and Ice Cream Company *
P.O. Box 119
Lawrence, KS
Territory: KS counties of Douglas, Johnson, Leavenworth, Franklin, Jefferson
Licensed: Milk, ice cream, cottage cheese
Date of Contract: 6/10/54
List of Licensees: 2/8/55, 3/15/55, 1/20/56, 6/20/56, 10/1/56, 2/25/57, 9/24/57, 2/14/58, 10/19/62

Lehigh Valley Cooperative Farmers
1000 North Seventh Street
Allentown, PA
Territory: NJ county of Warren; PA
counties of Lehigh, Northampton, Monroe,
Berks, Bucks, Montgomery, Carbon
Licensed: Milk
Date of Contract: 9/29/54
Termination Date: 11/30/57
List of Licensees: 2/8/55, 3/15/55,
1/20/56, 6/20/56, 10/1/56, 2/25/57,
9/24/57, 10/19/62

Lexington Dairy, Inc.
716 National Avenue
Lexington, KY
Territory: KY counties of Fayette, Scott,
Bourbon, Clark, Madison, Jessamine,
Woodford
Licensed: Milk, cottage cheese
Date of Contract: 9/20/54
List of Licensees: 2/8/55, 3/15/55,
10/19/62

Lindale Dairy Corporation*
124 W. Lexington Avenue
High Point, NC
Territory: NC counties of Guilford,
Forsyth, Caswell, Surry, Davie,
Rockingham, Stokes, Yadkin, Davidson,
Randolph
Licensed: Milk, ice cream, cottage
cheese
Date of Contract: 11/15/54
Termination Date: 12/31/57
List of Licensees: 2/8/55, 3/15/55,
1/20/56, 6/20/56, 10/1/56, 2/25/57,
9/24/57, 2/14/58, 10/19/62

Lindale Dairies, Inc.
Roanoke Rapids, NC
Territory: NC counties of Halifax,
Northampton, Bertie, Hertford, Warren,
Nash, Edgecombe; VA counties of
Brunswick, Greensville, Mecklenburg,
Southampton
Licensed: Milk
Date of Contract: 7/21/56
Termination Date: 9/30/59
List of Licensees: 10/1/56, 2/25/57,
9/24/57, 2/14/58, 10/19/62

Linton & Linton
180 E. Main Street
Wilmington, OH
Territory: OH counties of Clinton, Brown,
Greene, Warren, Highland, Clermont
Licensed: Milk, cottage cheese
Date of Contract: 1/1/54
Termination Date: 12/31/57
List of Licensees: 2/8/55, 3/15/55,
1/20/56, 6/20/56, 10/1/56, 10/19/62

Marin Dairymen's Milk Company.
Howard & 13th Street
San Francisco, CA
Territory: CA counties of San Francisco,
Santa Clara, San Mateo, Santa Cruz,
Marin
Licensed: Milk
Date of Contract: 4/16/51
List of Licensees: 11/15/51, 1/3/52,
7/10/52, 10/19/62

McClellan Dairies
352 Lincoln Avenue
Lancaster, OH
Territory: OH counties of Vinton, Jackson,
Galia, Fairfield, Pike, Lawrence, Morgan,
Meigs, Hocking, Perry, Athens, Scioto
Licensed: Milk, cottage cheese
Date of Contract: 12/21/54
Termination Date: 2/28/58
List of Licensees: 2/8/55, 3/15/55,
1/20/56, 6/20/56, 10/1/56, 2/25/57,
9/24/57, 2/14/58, 10/19/62

Med-O-Pure Dairy Foods *
Washington Court House, OH
Territory: OH counties of Fayette,
Madison, Ross, Adams, Union, Pickaway,
Highland, Clark, Clinton, Pike,
Champaign, Delaware, Greene, Fairfield,
Warren, Hocking
Licensed: Milk, ice cream
Date of Contract: 12/16/54
List of Licensees: 1/20/56, 6/20/56,
10/1/56, 2/25/57, 9/24/57, 2/14/58,
10/19/62

Medosweet Dairies, Inc.
2413 Pacific Avenue
Tacoma 1, WA
Territory: WA counties of Pierce, Thurston
Licensed: Milk
Date of Contract: 12/16/54
List of Licensees: 11/15/51, 1/3/52,
7/10/52, 10/19/62

Melville Dairy, Inc.
Burlington, NC
Territory: NC county of Alamance
Licensed: Milk
Date of Contract: 2/14/56
Termination Date: 3/31/59
List of Licensees: 10/1/56, 6/20/56,
2/25/57, 9/24/57, 2/14/58, 10/19/62

Minnesota Milk Company
370 University Avenue
St. Paul, MN
Territory: MN city of St. Paul
Licensed: Milk, cottage cheese
Date of Contract: 2/22/52
Termination Date: 6/9/57
List of Licensees: 7/10/52, 2/8/55,
3/15/55, 1/20/56, 10/1/56, 2/25/57,
10/19/62

Modesto College Dairy
Seventh & E Streets
Modesto, CA
Territory: CA city of Modesto
Licensed: Milk
Date of Contract: 11/1/51
List of Licensees: 11/15/51, 1/3/52,
7/10/52, 10/19/62

Nielsen's Creamery
147 South M Street
Tulare, CA
Territory: CA county of Tulare and cities of
Delano, Hanford, Corcoran, Kingsburg,
Selma, Reedley, Sanger, Ridgecrest
Licensed: Milk
Date of Contract: 8/1/51
List of Licensees: 11/15/51, 1/3/52,
7/10/52, 10/19/62

Northland Milk Company
East Sixth & Des Moines Avenue
Des Moines, IA
Territory: IA city of Des Moines
Licensed: Milk, cottage cheese
Date of Contract: 4/1/52
Termination Date: 3/31/57
List of Licensees: 7/10/52, 2/8/55,
3/15/55, 1/20/56, 6/20/56, 10/1/56,
2/25/57, 10/19/62

Northland Milk Company
11 West 28th Street
Minneapolis, MN
Licensed: Milk
Termination Date: 6/9/57
List of Licensees: 10/1/56, 2/25/57

Pevely Dairy Company
1001 S. Grand Avenue
St. Louis, MO
Territory: IL counties of Jersey, Madison,
St. Clair, Monroe, Randolph, Jackson,
Union, Pulaski; KY counties of Ballard,
McCracken; MO counties of St. Louis, St.

Charles, Jefferson, Perry, Washington, St.
Genevieve, Scott, Cape Girardeau, New
Madrid, Missouri
Licensed: Milk
Date of Contract: 12/1/54
List of Licensees: 2/8/55, 3/15/55,
10/19/62

Polk's Sanitary Milk *
1100 East 15th Street
Indianapolis, IN
Territory: IN counties of Howard, Tipton,
Clinton, Madison, Montgomery, Boone,
Hamilton, Putnam, Hendricks, Marion,
Hancock, Morgan, Johnson, Shelby, Owen,
Monroe, Brown, Bartholomew
Licensed: Milk, ice cream, cottage
cheese
Date of Contract: 12/6/54
Termination Date: 1/31/57
List of Licensees: 2/8/55, 1/20/56,
6/20/56, 10/1/56, 10/19/62

Producer's Dairy Delivery, Inc.
144 Belmont
Fresno, CA
Licensed: Milk, cottage cheese, orange
drink
List of Licensees: 11/15/51, 1/3/52,
7/10/52, 2/8/55, 3/15/55, 1/20/56,
6/20/56, 10/1/56, 2/25/57, 9/24/57,
12/1/58

Puritan Dairy Producers Company,
Inc. *
1217 North Broadway
Pittsburgh, KS
Territory: KS counties of Crawford,
Cherokee, Bourbon, Linn; MO counties of
Vernon, Jasper, Barton, Newton; OK
counties of Ottowa, Craig
Licensed: Milk, ice cream, cottage
cheese
Date of Contract: 8/15/54
List of Licensees: 2/8/55, 3/15/55,
1/20/56, 6/20/56, 10/1/56, 2/25/57,
9/24/57, 2/14/58, 10/19/62

Purity Maid Products Company *
1201 E. Market Street
New Albany, IN
Territory: IN counties of Floyd, Harrison,
Washington, Clark; KY counties of
Jefferson, Oldham
Licensed: Milk, ice cream, cottage
cheese
Date of Contract: 7/1/54
List of Licensees: 2/8/55, 3/15/55,
1/20/56, 6/20/56, 10/19/62

Roberts Dairy Company, Inc. *
4469 Farnum Street
Omaha, NE
Territory: Entire state of NE; SD counties
of Bonhomee, Charles Mix, Clay, Douglas,
Gregory, Hutcheson, Lincoln, Tripp,
Turner, Union, Yanktown; IA counties of
Audubon, Carroll, Buena Vista, Cass,
Cherokee, Clay, Crawford, Dickinson,
Fremont, Harrison, Idaho Lyon Mills,
Monona, Montgomery, O'Brien, Osceola,
Page, Plymouth, Pottawamie, Sac, Shelby,
Sioux, Woodbury
Licensed: Milk, ice cream, cottage
cheese
Date of Contract: 10/1/52
List of Licensees: 2/8/55, 3/15/55,
1/20/56, 10/19/62

Ross Dairy Corporation
451 River Street
Manistee, MI
Territory: MI counties of Manistee, Mason,
Lake, Wexford, Benzie, Oceania
Licensed: Milk
Date of Contract: 9/26/55
Termination Date: 11/30/58

List of Licensees: 1/20/56, 6/20/56,
10/1/56, 2/25/57, 9/24/57, 2/14/58,
10/19/62

Royal Crest Guernsey Farms, Inc. *
3977 Salem Avenue
Dayton 6, OH
Territory: OH counties of Montgomery,
Darke, Shelby, Miami, Preble, Butler,
Mercer, Clinton, Warren, Auglaize,
Champaign, Clark, Greene
Licensed: Milk, ice cream, cottage
cheese
Date of Contract: 12/7/54
List of Licensees: 2/8/55, 3/15/55,
1/20/56, 6/20/56, 10/1/56, 2/25/57,
9/24/57, 2/14/58, 10/19/62

Sanders Dairy *
1027 Washington Avenue
Piqua, OH
Territory: OH counties of Miami, Shelby,
Darke, Champaign, Clark, Auglaize
Licensed: Milk, ice cream, cottage
cheese
Date of Contract: 12/8/54
Termination Date: 1/31/58
List of Licensees: 2/8/55, 3/15/55,
1/20/56, 6/20/56, 10/1/56, 10/19/62

Sanitary Farm Dairies
133 F Street N.W.
Cedar Rapids, IA
Territory: IA counties of Benton, Johnson,
Cedar, Clinton, Scott, Washington, Iowa,
Jones, Linn, Tama; IL counties of
Whiteside, Henry, Mercer
Licensed: Milk, cottage cheese
Date of Contract: 11/1/53
List of Licensees: 2/8/55, 3/15/55,
10/19/62

Senn-Soldwedel Company
73 North First Avenue
Canton, IL
Territory: IL counties of Fulton,
McDonough, Schuyler, Knox, Warren
Licensed: Milk
Date of Contract: 8/1/57
List of Licensees: 9/24/57, 2/14/58,
10/19/62

Sorge Ice Cream & Dairy *
2000 South 10th Street
Manitowoc, WI
Territory: WI counties of Brown, Door,
Fond DuLac, Kewanee, Manitowoc,
Sheboygen, Ozaukee, Washington
Licensed: Milk, ice cream, cottage cheese
Date of Contract: 4/10/53
Termination Date: 4/9/57
List of Licensees: 2/8/55, 3/15/55,
1/20/56, 6/20/56, 10/1/56, 2/25/57,
9/24/57, 10/19/62

Sullivan Dairy Company
118 W. Jefferson
Sullivan, IL
Territory: IL counties of Douglas, Moultrie,
Coles, Piatt, Shelby, Cumberland, Jasper, ·
Edgar
Licensed: Milk
Date of Contract: 8/28/58
List of Licensees: 10/19/62

Sunrise Dairy, Inc. *
508 W. Franklin Avenue
Gastonia, NC
Territory: NC counties of Gaston,
Mecklenburg; SC county of York
Licensed: Milk, ice cream, cottage
cheese
Date of Contract: 1/3/55
Termination Date: 3/31/58
List of Licensees: 3/15/55, 1/20/56,
6/20/56, 10/1/56, 2/25/57, 9/24/57,
2/14/58, 10/19/62

Superior Dairies, Inc. *
P.O. Box 1509
Statesville, NC
*Territory: NC counties of Iredell,
Alexander*
Licensed: Milk, ice cream, cottage
cheese
Date of Contract: 1/3/55
Termination Date: 2/28/58
List of Licensees: 3/15/55, 1/20/56,
6/20/56, 10/1/56, 2/25/57, 2/14/58,
10/19/62

Teague Farms
RFD 1, Box 411
Montgomery, AL
*Territory: AL counties of Butler,
Montgomery, Crenshaw, Elmore,
Covington, Autauga*
Licensed: Milk
Date of Contract: 1/15/56
Termination Date: 2/28/59
List of Licensees: 10/1/56, 6/20/56,
2/25/57, 10/19/62

Ullery Dairy Company
Route 3
Greenville, OH
Territory: OH county of Darke
Licensed: Milk
Date of Contract: 3/20/55
Termination Date: 5/31/58
List of Licensees: 1/20/56, 6/20/56,
10/1/56, 2/25/57, 10/19/62

**Victory Dairy Company (Victory
Creamery, Inc.) ***
22 East Seventh Street
Emporia, KS
*Territory: KS counties of Lyon, Chase,
Greenwood, Osage, Coffee, Shawnee*
Licensed: Milk, ice cream, cottage
cheese
Date of Contract: 6/11/54
Termination Date: 8/31/57
List of Licensees: 2/8/55, 3/15/55,
1/20/56, 6/20/56, 10/1/56, 2/25/57,
9/24/ 2/14/58, 10/19/62

**Walla Walla Dairy Cooperative
Association**
44 South Palouse
Walla Walla, WA
Territory: WA city of Walla Walla
Licensed: Milk
Date of Contract: 12/1/51
List of Licensees: 11/15/51, 1/3/52,
7/10/52, 10/19/62

Wareham's Dairy, Inc.
1100 E. Adams Street
Taylorville, IL
*Territory: IL counties of Christian,
Montgomery, Macon, Logan, Shelby,
Macoupin, Snagamon, Effingham*
Licensed: Milk
Date of Contract: 8/28/58
List of Licensees: 10/19/62

H. B. Williams Dairy
3600 Telegraph Avenue
Oakland, CA
*Territory: CA counties of Alameda, Contra
Costa*
Licensed: Milk, cottage cheese, butter
Date of Contract: 11/1/51
List of Licensees: 11/15/51, 1/3/52,
7/10/52, 2/8/55, 10/19/62

Yakima Dairymen's Association
10 North 5th Avenue
Yakima, WA
Territory: WA city of Yakima
Licensed: Milk
Date of Contract: 12/1/51
List of Licensees: 1/3/52, 7/10/52,
10/19/62

Yakima Valley Dairy
Yakima, WA
Licensed: Milk
List of Licensees: 11/15/51

ICE CREAM: These are ice cream franchises only - no other milk products.

Beechmont Dairy, Inc.
2710 North Avenue
Bridgeport, CT
Territory: CT county of Fairfield
Date of Contract: 4/27/53
List of Licensees: 10/19/62

Boyles Dairy & Ice Cream
Ruleville, MS
*Territory: MS counties of Bolivar,
Coahoma, Leflore, Quitman, Sunflower,
Tallachatchai, Washington*
Date of Contract: 5/8/53
List of Licensees: 10/19/62

Carnation Company
5045 Wilshire Boulevard
Los Angeles 36, CA
*Territory: Entire states of AZ and CA; NV
counties of Washoe, Storey, Lyon, Douglas,
Ormsby, Pershing, Churchill, Mineral*
Date of Contract: 9/22/50
List of Licensees: 11/15/51, 10/19/62

Cresthaven Dairies, Inc.
721 Pillsbury Street
St. Paul, MN
*Territory: Entire state of MN; IA counties
of Worth, Mitchell, Winnebago, Howard,
Winneshiek; ND counties of Cass,
Richland, Trail; SD counties of Deuel,
Brookings, Codington, Moody, Minnehaha,
Roberts*
Date of Contract: 1/14/53
List of Licensees: 10/19/62

Elizabethtown Ice Cream Company
South Main Street
Elizabethtown, KY
*Territory: KY counties of Spencer, Bullitt,
Meade, Hardin, Nelson, Washington,
Marion, Larue, Grayson, Hart, Green,
Taylor, Adair, Metclafe, Cumberland,
Monroe*
Date of Contract: 3/20/55
List of Licensees: 1/20/56, 9/24/57,
10/19/62

Golden Locks Ice Cream Company
221 E. Ransom Street
Kalamazoo, MI
*Territory: MI counties of Kalamazoo,
Calhoun, Branch, St. Joseph, Cass, Berrien,
Van Buren, Barry, Easton, Allegan,
Jackson, Hillside*
Date of Contract: 7/1/57
List of Licensees: 9/24/57, 10/19/62

Hannibal Quality Dairy, Inc.
Hannibal, MO
List of Licensees: 9/24/57

Henderson Creamery Company, Inc.
120 First Street West
Henderson, KY
Territory: KY county of Henderson
Date of Contract: 6/1/54
List of Licensees: 2/8/55, 3/15/55,
1/20/56, 10/19/62

Hudson's
Fulton, NY
*Territory: NY counties of Oswego,
Wayne*
Date of Contract: 6/3/53
List of Licensees: 10/19/62

Irwin Ice Cream Company
Port Huron, MI
*Territory: MI counties of St. Clair, Sanilac,
Macomb*
Date of Contract: 3/19/53
List of Licensees: 10/19/62

LaBelle Creamery
Belle Fourche, SD
*Territory: SD counties of Butte, Meade,
Lawrence, Harding, Custer, Pennington,
Fall River; MT county of Carter; WY
county of Crook*
Date of Contract: 7/6/53
List of Licensees: 2/8/55, 3/15/55,
1/20/56, 10/19/62

Land O'Sun Dairies, Inc.
101 Alton Road
Miami Beach, FL
*Territory: FL counties of Monroe, Dade,
Broward, Martin, Lee, Palm Beach, Collier,
Hendry, Charlotte, Sarasota, Manatee,
Hillsborough, Pinellas, Polk, Glades,
Desota, Hardee*
Date of Contract: 3/20/55
List of Licensees: 1/20/56, 9/24/57,
10/19/62

Luick Ice Cream Company
601 East Ogden Avenue
Milwaukee 1, WI
Territory: Entire state of MI
Date of Contract: 8/21/50
List of Licensees: 11/15/51, 10/19/62

George Moore Ice Cream Company
54 Alabama Street S.W.
Atlanta, GA
*Territory: All GA counties north of and
including Troup, Pike, Meriweather, Lamar,
Monroe, Bibb, Jones, Baldwin, Hancock,
Warren, McDuffie, Columbia*
Date of Contract: 9/10/53
List of Licensees: 10/19/62

**Mt. Vernon Ice Cream & Bottling
Company**
12th and Broadway
Mt. Vernon, IL
*Territory: IL counties of Jefferson,
Washington, Franklin, Wayne, Hamilton,
Marion, Richland, Montgomery,
Macoupin, Fayette, Effingham*
Date of Contract: 12/1/54
List of Licensees: 2/8/55, 3/15/55,
1/20/56, 9/24/57, 10/19/62

O'Fallon Quality Dairy Company
Oak Street & Highway 50
O'Fallon, IL
*Territory: IL counties of Bond, Clinton,
Madison, St. Clair*
Date of Contract: 3/6/53
List of Licensees: 2/8/55, 3/15/55,
1/20/56, 9/24/57, 10/19/62

Paradise Ice Cream Company
Orangeburg, SC
*Territory: SC counties of Bamberg,
Calhoun, Grangeburg, Colleton, Richland,
Lexington, Barnwell, Hampton,
Dorchester, Aiken*
Date of Contract: 7/1/56
List of Licensees: 9/24/57, 10/19/62

Premium Dairy Foods, Inc.
217 North 6th Street
Independence, KS
*Territory: KS counties of Allen, Linn,
Chautauqua, Anderson, Neosho,
Montgomery, Wilson, Woodson, Elk,
Bourbon, Crawford, Labette, Coffey,
Greenwood*
Date of Contract: 2/17/53
List of Licensees: 2/8/55, 3/15/55,
1/20/56, 10/19/62

Richards Dairy
205 Brush Street
St. Johns, MI
*Territory: MI counties of Clinton, Ionia,
Gratiot*
Date of Contract: 3/21/53
List of Licensees: 10/19/62

Richmond Ice Cream Company
Third and Broadway
Richmond, KY
*Territory: KY counties of Boyle, Clark,
Clay, Estill, Fayette, Garrard, Jessamine,
Jackson, Lee, Laurel, Lincoln, Madison,
Mercer, Owsley, Rockcastle, Woodford*
Date of Contract: 5/1/54
List of Licensees: 2/8/55, 3/15/55,
1/20/56, 10/19/62

Ringold Ice Cream Company, Inc.
Chillicothe, OH
*Territory: OH counties of Ross, Pickaway,
Fayette, Madison, Union, Adams,
Highland, Fairfield, Hocking, Jackson,
Lawrence, Morgan, Pike, Vinton, Meigs,
Galia, Athens, Muskingum, Perry*
Date of Contract: 12/16/54
List of Licensees: 9/24/57, 10/19/62

MISCELLANEOUS FOODS:

American Home Foods, Inc.
22 East 40th Street
New York, NY
Licensed: Puddings, pie fillings, gelatin
desserts
Date of Contract: 3/25/52
Termination Date: 4/1/55
List of Licensees: 1/3/52, 7/10/52,
10/19/62

Bol Manufacturing Company
5145 Lake Street
Chicago, IL
Licensed: Hopalong Aid, assorted
flavors powdered non-carbonated fruit
concentrates
Date of Contract: 8/25/50
Termination Date: 4/1/55
List of Licensees: 11/15/51, 1/3/52,
7/10/52, 10/19/62

R. D. Bradshaw & Sons
Wendell, ID
Licensed: Spunny Spread, honey with
other flavors, packaged in glass and/or
other containers
Date of Contract: 4/29/52
List of Licensees: 7/10/52, 4/1/55,
10/19/62

Cella Vineyards
P.O. Box 312
Fresno, CA
Licensed: Bottled and canned Betsy
Ross grape juice
Date of Contract: 6/10/52
Termination Date: 4/1/55
List of Licensees: 7/10/52, 2/8/55, 10/19/62

Gordon Foods, Inc.
1075 Sylvan Road, S.W.
Atlanta, GA
Licensed: Potato chips
List of Licensees: 11/15/51

Frederick W. Huber, Inc.
268 West Broadway
New York, NY
Licensed: Jams, jellies, marmalades,
preserves
Date of Contract: 2/21/52
Termination Date: 4/1/55
List of Licensees: 7/10/52, 10/19/62

Kuehmann Foods, Inc.
1228 Oakwood Avenue
Toledo, OH
Licensed: Potato chips, shoestring
potatoes
Date of Contract: 7/10/50
Termination Date: 4/1/55
List of Licensees: 11/15/51, 1/3/52,
10/19/62

Lady's Choice Foods
4578 Worth Street
Los Angeles 33, CA
Licensed: Peanut butter, jams, jellies,
honey
Date of Contract: 8/16/50
Termination Date: 4/1/55
List of Licensees: 11/15/51, 1/3/52,
7/10/52, 10/19/62

J. Langrall & Brothers, Inc.
2105 Aliceanna Street
Baltimore 31, MD
Licensed: Canned green beans
Date of Contract: 2/6/52
Termination Date: 4/1/55
List of Licensees: 7/10/52, 10/19/62

Marlo Packing Company
1955 Carroll Avenue
San Francisco, CA
Licensed: Canned beef stew, spaghetti
& meatballs, chili con carne
Date of Contract: 3/2/53
Termination Date: 4/1/55
List of Licensees: 7/10/52, 10/19/62

**Maryland Popcorn Cooperative
Association, Inc.**
Riderwood, MD
Licensed: Hopalong Cassidy's Favorite
Popcorn
Date of Contract: 3/2/53
Termination Date: 4/1/55
List of Licensees: 10/19/62

Northwest Cone Company
1400 West 37th Street
Chicago, IL
Licensed: Ice cream cones, cups,
packaged straws
Date of Contract: 9/15/52
List of Licensees: 10/19/62

Rome Frozen Food Lockers
709 W. Dominick Road
Rome, NY
Licensed: Laminated frozen beef steaks
Date of Contract: 2/5/52
Termination Date: 4/1/55
List of Licensees: 7/10/52, 10/19/62

Ryan Products, Inc.
420 Lexington Avenue
New York 17, NY
Licensed: Chocolate candy bar

Scenic Citrus Cooperative
Frostproof, FL
Licensed: Single strength canned juices
including orange, grapefruit, tangerine,
lime,and orange/grapefruit blend;
pasteurized concentrated juices
including orange, grapefruit, tangerine
and orange/grapefruit blend
Date of Contract: 9/10/51
Termination Date: 4/1/55
List of Licensees: 11/15/51, 1/3/52,
7/10/52, 10/19/62

Topps Chewing Gum Company, Inc.
237 West 37th Street
Brooklyn, NY
Licensed: Hard candy, taffy, chewing
gum
Date of Contract: 7/1/50
*List of Licensees: 11/15/51, 1/3/52, 7/10/52,
10/19/62*

Van Camp Sea Foods, Inc.
772 Tuna Street
Terminal Island, CA
Licensed: Chicken-Of-The-Sea tuna,
other sea food products
Date of Contract: 6/17/52
Termination Date: 4/1/55
List of Licensees: 7/10/52, 10/19/62

W. T. Young Foods, Inc.
767 E. Third Street
Lexington, KY
Licensed: Peanut butter
Date of Contract: 3/1/51
Termination Date: 4/1/55
List of Licensees: 11/15/51, 1/3/52,
7/10/52, 2/8/55, 10/19/62

MERCHANDISE

CLOTHING AND
ACCESSORIES:

**Acme Boot Manufacturing Company,
Inc.**
Clarksville, TN
Licensed: Children's boots, slippers
Date of Contract: 9/1/50
List of Licensees: 8/1/50, 11/15/51,
2/1/52, 7/10/52, Pocket Guide of
Hopalong Cassidy Western Wear No. 4
Spring 1953, 5/29/54, 10/19/62

Eli E. Albert, Inc.
Albert Building
159 Fifth Avenue
New York 10, NY
Licensed: Jodphur ensembles, slip-
overs, jackets, Eton caps, black
gabardine unlined zippered jacket with
matching boxer longee
Date of Contract: 4/29/49
List of Licensees: 8/1/50, 11/15/51,
2/1/52, 7/10/52, Pocket Guide of
Hopalong Cassidy Western Wear No. 4
Spring 1953, 5/19/54

[Eli E. Albert, Inc.]
935 Broadway
New York, NY
List of Licensees: 2/1/56, 6/20/56,
10/19/62

Anson, Inc.
24 Baker Street
Providence 5, RI
Licensed: Boys' jewelry and accessories
including chains, cuff links, money
clips, tie holders
Date of Contract: 7/1/50
List of Licensees: 8/1/50, 11/15/51,
2/1/52, 7/10/52, 10/19/62

Art Gloves
320 W. Ohio Street
Chicago 10, IL
Licensed: gloves
List of Licensees: 8/1/50, 2/1/52,
Pocket Guide of Hopalong Cassidy
Western Wear No. 4 Spring 1953
See Also: Hugger Gloves

George S. Bailey Hat Company, Inc.
[Bailey of Hollywood]
716 S. Los Angeles Street
Los Angeles 14, CA
(New York Office: 358 Fifth Avenue)
Licensed: Boys' and girls' wool felt
cowboy hats (Deputy, Hoppy Jr, and
Topper)
Date of Contract: 6/15/51
List of Licensees: 8/1/50, 11/15/51,
2/1/52, 7/10/52, Pocket Guide of
Hopalong Cassidy Western Wear No. 4
Spring 1953, 5/19/54, 6/20/56,
10/23/57, 4/10/58, 12/1/58, 10/19/62

Barclay Knitwear Company
1239 Broadway
New York, NY
Licensed: Jacquards and sweaters
including plain knit wool, and/or
synthetic fibre, and/or cotton, long
sleeve, short sleeve, and sleeveless
Date of Contract: 7/1/50
List of Licensees: 8/1/50, 11/15/51,
2/1/52, 7/10/52, Pocket Guide of
Hopalong Cassidy Western Wear No. 4
Spring 1953, 10/19/62

Blue Bell, Inc.
93 Worth Street
New York 13, NY
Licensed: Boys' and girls' cotton denim
jeans and dungarees, jackets, boys'
denim and jean shirts
Date of Contract: 7/1/50
List of Licensees: 8/1/50, 11/15/51,
2/1/52, 7/10/52, 5/19/54, 2/1/56,
6/20/56, 10/23/57, 12/1/58, 10/19/62

[Blue Bell, Inc.]
350 Fifth Avenue
New York 1, NY
Licensed: Pocket Guide of Hopalong
Cassidy Western Wear No. 4 Spring
1953

Broadworth Sales Corporation
93 Worth Street
New York 13, NY
Licensed: Children's knit underwear
Date of Contract: 10/13/50
List of Licensees: 11/15/51, 7/10/52

[Broadworth Mills, Inc.]
350 Fifth Avenue
New York 1, NY
List of Licensees: 2/1/52, Pocket Guide
of Hopalong Cassidy Western Wear No.
4 Spring 1953, 5/19/54, 10/19/62

Burlington Mills Corporation
Ribbon Division
350 Fifth Avenue
New York, NY
Licensed: Girls' hair bow ribbon
ensembles
Date of Contract: 7/26/50
List of Licensees: 11/15/51, 10/19/62

Cable Raincoat Company
72 Northampton Street
Boston 18, MA
Licensed: Boys' and girls' raincoats and
rain hats
Date of Contract: 3/8/51
List of Licensees: 11/15/51, 2/1/52,
7/10/52, 10/19/62

California Sportswear Company
1024 Maple Street
Los Angeles, CA
Licensed: Boys' Western style leather
jackets
Date of Contract: 7/1/50
List of Licensees: 8/1/50, 11/15/51,
2/1/52, 10/19/62

Camco Sportswear, Inc.
15 Ann Street
South Norwalk, CT
Licensed: Boys' and girls' swim suits
Date of Contract: 8/13/51
List of Licensees: 11/15/51, 2/1/52,
10/19/62

Coro, Inc.
167 Point Street
Providence 2, RI
Licensed: Girls' jewelry and accessories
Date of Contract: 7/1/50
List of Licensees: 8/1/50, 2/1/52,
Pocket Guide of Hopalong Cassidy

Western Wear No. 4 Spring 1953,
5/19/54, 2/1/56, 6/20/56, 10/23/57,
4/10/58, 12/1/58, 10/19/62

[Coro, Inc.]
47 West 34th Street
New York, NY
List of Licensees: 11/15/51, 7/10/52

J. T. Flagg Knitting Company, Inc.
Florence, AL
Licensed: Children's sweatshirts sizes
2-20, T shirts, polo shirts
Date of Contract: 7/1/50
List of Licensees: 8/1/50, 11/15/51,
2/1/52, 7/10/52, 5/19/54

[Flagg-Utica Corporation]
Florence, AL
List of Licensees: Pocket Guide of
Hopalong Cassidy Western Wear No. 4
Spring 1953, 2/1/56, 6/20/56,
10/23/57, 10/19/62

John A. Frye Shoe Company
Marlborough, MA
Licensed: leather boots
List of Licensees: 8/1/50

Gary's Sportswear of California
239 S. Los Angeles Street
Los Angeles, CA
Licensed: Western dress jackets
List of Licensees: 8/1/50

Ernest Glick Company
160 North Wacker Drive
Chicago 6, IL
Licensed: Boys' and girls' neckwear,
scarves, handkerchiefs, tie conchos, tie
rings, tie racks, brush holders
Date of Contract: 8/1/50
List of Licensees: 8/1/50, 11/15/51,
2/1/52, 7/10/52, Pocket Guide of
Hopalong Cassidy Western Wear No. 4
Spring 1953, 5/19/54, 2/1/56,
6/20/56, 10/23/57, 4/10/58, 12/1/58

[Ernest Glick Company]
315 West Adams Street
Chicago, IL
List of Licensees: 10/19/62

Gold Seal Rubber Company
174 Lincoln Street
Boston, MA
Licensed: Cowboy style boots with
Hopalong Cassidy decoration (junior's
sizes 13-3, children's sizes 5-12)
Licensed: Pocket Guide of Hopalong
Cassidy Western Wear No. 4 Spring
1953

Goodyear Rubber Company
Middletown, CT
Licensed: Rubber cowboy boots, tennis
shoes
Date of Contract: 10/4/49
List of Licensees: 8/1/50, 11/15/51,
2/1/52, 7/10/52, 5/19/54, 2/1/56,
6/20/56, 10/23/56, 4/10/58, 12/1/58,
10/19/62

His Nibs Shirt Corporation
Souderton, PA
Licensed: Boys' and girls' Western style
shirts, boys' pajamas
Date of Contract: 12/16/53
List of Licensees: 5/19/54, 2/1/56,
6/20/56, 10/23/57, 4/10/58, 10/19/62

Hugger Gloves
Division of Portis Style Industries, Inc.
320 West Ohio Street
Chicago 10, IL
(Sales Offices: 1199 Broadway, NY 1; 9
North 3rd Street, Philadelphia 6, PA)
Licensed: Lined and unlined gauntlet

cape glove, children's wool worsted mittens and gloves
Licensed: Pocket Guide of Hopalong Cassidy Western Wear No. 4 Spring 1953
See Also: *Art Gloves*

Herman Iskin & Company, Inc.
Souderton, PA
Licensed: Boys' and girls' cowboy and cowgirl playsuit ensembles consisting of collapsible cotton hat, pants or skirt, vest, shirt, kerchief, lariat, belt holster or pistol; chaps and vest ensemble and skirt and vest ensemble; boys' boxed set ensemble consisting of pair of boxer-type longees, pair of boxer-type shorts and shirt
List of Licensees: 8/1/50, 11/15/51, 7/10/52

[Herman Iskin & Company, Inc.]
Telford, PA
Licensed: Pocket Guide of Hopalong Cassidy Western Wear No. 4 Spring 1953

[Herman Iskin & Company, Inc.]
200 Fifth Avenue
New York 10, NY
List of Licensees: 2/1/52, Pocket Guide of Hopalong Cassidy Western Wear No. 4 Spring 1953, 5/19/54, 2/1/56, 6/20/56, 10/23/57, 4/10/58, 12/1/58, 10/19/62

J Bar T, Inc.
1620 S. Flower Street
Los Angeles 15, CA
Licensed: Boys' and girls' Western style frontier suit ensembles including girls' Western shirt and skirt, girls' Western dress suit with jacket and skirt, girls' dress jacket
Date of Contract: 7/1/50
List of Licensees: 8/1/50, 11/15/51, 2/1/52, 7/10/52, Pocket Guide of Hopalong Cassidy Western Wear No. 4 Spring 1953, 5/19/54, 10/19/62

Morton Karten, Inc.
6101 Sixteenth Avenue
Brooklyn, NY
Licensed: Snowsuits
List of Licensees: 8/1/50

Klinkerfues Manufacturing Company
895-907 E. Seventh Street
St. Paul 6, MN
Licensed: Boys' fabric jackets and storm coats, sizes 4-12
Date of Contract: 9/1/50
List of Licensees: 11/15/51, 2/1/52, 10/19/62

Little Champ of Hollywood
730 S. Los Angeles Street
Los Angeles 14, CA
Licensed: Boys' Western style shirts, sports shirts and cowboy shirts in rayon poplin and rayon gabardine
Date of Contract: 3/1/51
List of Licensees: 8/1/50, 11/15/51, 2/1/52, 7/10/52, Pocket Guide of Hopalong Cassidy Western Wear No. 4 Spring 1953, 10/19/62

Joseph Love, Inc.
1333 Broadway
New York, NY
Licensed: Girls' skirts and ensembles
List of Licensees: 8/1/50

Morse and Morse, Inc.
240 South Broadway
Los Angeles, CA
Licensed: Children's and adults' T

shirts and polo shirts, children's knitted pajamas
Date of Contract: 4/11/49
List of Licensees: 8/1/50, 11/15/51, 2/1/52, 7/10/52, 10/19/62

Norwich Mills, Inc.
Norwich, NY
Licensed: Boys' and girls' T shirts, sweatshirts, polo shirts
Date of Contract: 4/1/58
List of Licensees: 4/10/58, 12/1/58, 10/19/62

Oxford Boyswear, Inc.
16 West 19th Street
New York, NY
Licensed: Boys' Western frontier or stockman type dress pants and slacks
Date of Contract: 6/28/49
List of Licensees: 8/1/50, 11/15/51, 2/1/52, 7/10/52, 10/19/62

Lurie Pizer Company
943 S. Wall Street
Los Angeles, CA
Licensed: Girls' ensembles
List of Licensees: 8/1/50

Pied Piper of Dallas
918 W. Commerce Street
Dallas 8, TX
Licensed: Boys' and girls' Western style ensembles including girls' skirt and bolero, boys' and girls' boxer shorts, jacket and suspender jean or boxer sets, boy's jacket and jean sets
List of Licensees: 8/1/50
See Also: *Walls of Texas, Inc.*

[Pied Piper, Inc.]
Cleburne, TX
List of Licensees: 11/15/51, 2/1/52, 7/10/52

Portis Style Industries, Inc.
320 West Ohio Street
Chicago, IL
Licensed: Boys' and girls' gloves and mittens
Date of Contract: 7/1/50
List of Licensees: 11/15/51, 7/10/52, 5/19/54, 2/1/56, 6/20/56, 10/19/62

Princess Pat Panty Company
45 West 34th Street
New York, NY
Licensed: Two and three piece cotton knit ensembles consisting of cardigan, slipover, and/or pants, sizes 1-6x
Date of Contract: 1/16/51
List of Licensees: 11/15/51, 2/1/52, 10/19/62

George Russell Sportswear
716 S. Los Angeles Street
Los Angeles, CA
Licensed: Boys' cotton poplin and cotton gabardine jackets with zipper front
List of Licensees: 8/1/50, 11/15/51

Spatz Brothers, Inc.
1141 Broadway
New York, NY
Licensed: Rainwear
List of Licensees: 8/1/50

Sport-Wear Hosiery Mills, Inc.
350 Fifth Avenue
New York 1, NY
Licensed: Boys' and girls' slipper socks with leatherette soles, boys' and girls' hosiery, men's and women's slipper socks, dental kit
Date of Contract: 7/1/50
List of Licensees: 8/1/50, 11/15/51,

2/1/52, 7/10/52, Pocket Guide of Hopalong Cassidy Western Wear No. 4 Spring 1953, 5/19/54, 2/1/56, 10/19/62

Van Baalen-Heilbrun and Company, Inc.
1237-39 Broadway
New York, NY
Licensed: Robes
List of Licensees: 8/1/50

Walls of Texas, Inc./for Pied Piper, Inc.
Cleburne, TX
Licensed: Boys' and girls' Western style ensembles including girls' skirt and bolero, boys' and girls' boxer shorts, jacket and suspender jean or boxer sets, boy's jacket and jean sets
Date of Contract: 9/1/51
List of Licensees: Pocket Guide of Hopalong Cassidy Western Wear No. 4 Spring 1953, 5/19/54, 10/19/62
See Also: *Pied Piper, Inc., Pied Piper of Dallas*

Weldon Manufacturing Company
1270 Broadway
New York, NY
Licensed: Pajamas
List of Licensees: 8/1/50

Yale Belt Corporation
183 Wooster Street
New York 12, NY
Licensed: Children's leather belts, elastic suspenders, Leathercraft Kits
Date of Contract: 5/1/49
List of Licensees: 8/1/50, 2/1/52, 7/10/52, Pocket Guide of Hopalong Cassidy Western Wear No. 4 Spring 1953, 5/19/54, 10/19/62

DOMESTICS:

Bell Textile Company, Inc.
353 Broadway
New York 13, NY
Licensed: Children's chenille bedspreads, bedroom rugs, draperies, bath mats and lids
Date of Contract: 7/1/50
List of Licensees: 11/15/51, 2/1/52, 7/10/52, Pocket Guide of Hopalong Cassidy Western Wear No. 4 Spring 1953, 5/19/54, 2/1/56, 6/20/56, 10/23/57, 4/10/58

[Bell Textile Company, Inc.]
295 Fifth Avenue
New York 16, NY
List of Licensees: 12/1/58, 10/19/62

R. A. Briggs & Company
1500 S. Laflin Street
Chicago, IL
Licensed: Children's towels
Date of Contract: 1/30/51
List of Licensees: 11/15/51, 2/1/52, 7/10/52, 10/19/62

Columbia Mills, Inc.
428 S. Warren Street
Syracuse, NY
Licensed: Western style window shades
Date of Contract: 7/1/50
List of Licensees: 10/19/62

Paulsboro Manufacturing Company
Architects Building
17th and Sansom Streets
Philadelphia 3, PA
Licensed: Linoleum floor coverings and wainscoting
Date of Contract: 12/18/50
List of Licensees: 11/15/51, 2/1/52,

7/10/5, Pocket Guide of Hopalong Cassidy Western Wear No. 4 Spring 1953, 5/19/54, 10/19/62

Peter Pan Products
22 W. Monroe Street
Chicago 3, IL
Licensed: Cookie jar, cake molds

Star Embroidery Company
719 S. Los Angeles Street
Los Angeles, CA
Licensed: Wash cloths and towels

The Ullman Company, Inc.
319 McKibbin Street
Brooklyn 6, NY
Licensed: Plastic or cork table mats
Date of Contract: 7/1/50
List of Licensees: 11/15/51, 2/1/52, 7/10/52, Pocket Guide of Hopalong Cassidy Western Wear No. 4 Spring 1953, 5/19/54, 2/1/56, 6/10/56, 10/19/62

Vita-Cress Company
P.O. Box 761
Mesa, AZ
Licensed: Chuckwagon garden sets, boxed, refills consisting of 3/4 oz Vita-Cress seeds in cloth bags
Date of Contract: 7/5/51
List of Licensees: 11/15/51, 2/1/52, 7/10/52
See Also: *Southwest Traders, Inc.*

Wilmart Products Corporation
9 West 18th Street
New York, NY
Licensed: Laundry, shoe and clothes bags
Date of Contract: 1/8/51
List of Licensees: 11/15/51, 7/10/52

[Wilmart Products Corporation]
30 East 21st Street
New York 10, NY
List of Licensees: 2/1/52, Pocket Guide of Hopalong Cassidy Western Wear No. 4 Spring 1953, 10/19/62

FURNITURE

Aladdin Industries, Inc.
703 Murfreesboro Road
Nashville, TN
Licensed: Lamps
Date of Contract: 7/1/50
List of Licensees: 11/15/51, 2/1/52, 7/10/52, Pocket Guide of Hopalong Cassidy Western Wear No. 4 Spring 1953, 5/19/54, 2/1/56, 6/20/56, 10/23/57, 4/10/58, 12/1/58, 10/19/62

Comfort Lines, Inc./for Triple A Products
1735 Diversey Boulevard
Chicago 14, IL
Licensed: Hopalong Cassidy comfy chair chrome chairs
List of Licensees: 10/19/62
See Also: *Triple A Products*

Loroman Company
1150 Broadway
New York 1, NY
Licensed: Metal or fibre clothes hampers
List of Licensees: 11/15/51, 2/1/52

Morgan Manufacturing Company
Ashville, NC
Licensed: Children's bedroom furniture including beds, dressers, clothes closets, vanity cases, chests of drawers, combination desk/chest of drawers, chairs, horse rocker chairs, regular

chairs and tables
Date of Contract: 7/1/50
List of Licensees: 11/15/51, 7/10/52

[Morgan Manufacturing Company]
P.O. Box 458
Black Mountain, NC
List of Licensees: 10/19/62

Page Distributors
3031 West Vernon
Los Angeles, CA
Licensed: Children's contour Hoppy
Lounge Chair
Date of Contract: 2/15/53
List of Licensees: Pocket Guide of
Hopalong Cassidy Western Wear No. 4
Spring 1953, 10/19/62

Ramorse, Inc.
Fitchburg, MA
Licensed: Ranch style wooden toy
chests
Date of Contract: 4/15/51
List of Licensees: 11/15/51, 2/1/52,
7/10/52, 10/19/62

Triple A Products
1735 Diversey Boulevard
Chicago 14, IL
Licensed: Hopalong Cassidy comfy
chair chrome chairs
See Also: *Comfort Lines, Inc.*

GLASS AND DINNER-WARE

W. S. George Pottery Company
East Palestine, OH
Licensed: Children's semi-vitreous
dinnerware including plate, bowl, mug
and/or cup
Date of Contract: 7/1/50
List of Licensees: 11/15/51, 2/1/52,
7/10/52, Pocket Guide of Hopalong
Cassidy Western Wear No. 4 Spring
1953, 5/19/54, 2/1/56, 10/23/57, 4/
10/58, 12/1/58, 10/19/62

Samuel Mallinger & Company
Forbes & Gist Streets
Pittsburgh 19, PA
Licensed: Children's glass cereal bowls,
water and juice glasses, milk glass
tumblers
Date of Contract: 7/1/50
List of Licensees: 11/15/51, 2/1/52,
7/10/52, 10/19/62

C. A. Reed Company
Williamsport, PA
Licensed: Paper dinnerware including
matching party plates, napkins,
tablecloths, and cups, paper nut cups
Date of Contract: 7/1/50
List of Licensees: 11/15/51, 2/1/52,
7/10/52, Pocket Guide of Hopalong
Cassidy Western Wear No. 4 Spring
1953, 5/19/54, 2/1/56, 6/20/56,
10/19/62

PROMOTIONAL AND SUPPLIES

Bloomer Brothers Company
Newark, NY

Burd & Fletcher Company
Seventh Street, May to Central
Kansas City 6, MO

Bruce Carton Company
3400 Chelsea Avenue
Memphis 8, TN

Color Laboratories, Inc.
9216 S. E. Ramona Street
Portland 66, OR
Licensed: Color film and prints for
photo contests
Date of Contract: 6/24/51
List of Licensees: 11/15/51, 7/10/52,
10/19/62

Container Corporation of America
1301 West 35th Street
Chicago 9, IL

Continental Can Company, Inc.
100 East 42nd Street
New York 17, NY

Continental Lithographers, Inc.
21 S. Ninth Street
St. Louis 2, MO
Licensed: 24-sheet billboard posters for
authorized bread and dairy licensees
List of Licensees: 11/15/51, 7/10/52

Crouch Dairy Supply Company, Inc.
321 South Main Street
Fort Worth, TX

Dixie Cup Company
Easton, PA

Fireboard Paper Products Corporation
475 Brannan Street
San Francisco 19, CA

Gaylord Container Corporation
Division of Crown Zellerbach
Corporation
3000 South Second Street
St. Louis 18, MO

Golden State Sales Corporation
425 Battery Street
San Francisco 11, CA

Hopalong Cassidy Enterprises
8907 Wilshire Boulevard
Beverly Hills, CA
Licensed: Hoppy and Topper blow-ups
(sizes 17" x 22", 24" x 36"), Hoppy TV
sound films (15 second and 50 second
lengths), newspaper mats
List of Licensees: Pocket Guide of
Hopalong Cassidy Western Wear No. 4
Spring 1953

Industrial Paper Company
Single Service Division
220 East 42nd Street
New York 17, NY

**Lehmann Printing & Lithographing
Company**
300 Second Street
San Francisco, CA

Lily-Tulip Cup Corporation
122 East 42nd Street
New York 17, NY

Joe Lowe Corporation
601 West 26th Street
New York 1, NY

Marathon Paper Company
A Division of American Can Company
Menasha, WI
Licensed: End labels, bread wrappers,
outserts
List of Licensees: 11/15/51, 7/10/52

Milprint, Inc.
431 W. Florida Street
Milwaukee, WI

Licensed: Cellophane bread wrappings
List of Licensees: 11/15/51, 7/10/52

Mold Craft Company
3727 N. Palmer Street
Milwaukee, WI
Licensed: Three dimensional rubber
and/or plastic molded and/or cast
Western figures, both full- round and
half-round for display purposes only
Date of Contract: 3/22/51
List of Licensees: 7/10/52, 10/19/62

Owens-Illinois Glass Company
320 California Street
San Francisco 19, CA
Licensed: Glass milk bottles, glass dairy
tumblers, glass cheese containers
List of Licensees: 11/15/51, 7/10/52

[Owens-Illinois Glass Company]
Toledo 1, OH

Pacific Waxed Paper Company
1050 Sixth Avenue, South
Seattle 4, WA
Licensed: End labels, bread wrappers,
outserts and/or inserts
List of Licensees: 11/15/51, 7/10/52

Plastic Container Corporation
West Warren, MA

Pollack Paper Corporation
Dallas, TX
Licensed: End labels, bread wrappers,
outserts and/or inserts
List of Licensees: 11/15/51, 7/10/52

Premium Plastics Company, Inc.
22 West Monroe Street
Chicago, IL

Sealright Corporation
Fulton, NY

Standard Packaging Corporation
1200 West Fullertown Avenue
Chicago 14, IL

W. L. Stensgaard and Associates
346 N. Justine Street
Chicago, IL
Licensed: Promotional displays
including mechanical displays, full
dimensional figures, flat cutouts,
window strips or streamers, silk screen
posters of "comuras" (commercial
murals)
List of Licensees: 11/15/51

**Thatcher Glass Manufacturing
Company, Inc.**
Elmira, NY
Licensed: Glass milk bottles, glass food
product containers
List of Licensees: 11/15/51, 7/10/52

Western Novelty Company
P.O. Box 285
North Hollywood, CA
Licensed: Giveaways
List of Licensees: 2/1/52, Pocket Guide
of Hopalong Cassidy Western Wear No.
4 Spring 1953

Western Waxed Paper Division
Crown Zellerbach Corporation
5900 Sheila Street
East Los Angeles, CA
Licensed: End labels, bread wrappers,
outserts and/or inserts
List of Licensees: 11/15/51, 7/10/52

STATIONERY:

**Art Leather Manufacturing Company,
Inc.**
10-34 Forty-fourth Drive
Long Island City, NY
Licensed: Photo mounts and picture
frames
Date of Contract: 1/18/51
List of Licensees: 11/15/51, 2/1/52,
10/19/62

Banthrico Industries, Inc.
17 N. Des Plaines Street
Chicago 6, Il
Licensed: Hopalong Cassidy Bank
Savings Plan for Children
Date of Contract: 7/1/50
List of Licensees: 11/15/51, 7/10/52,
10/19/62

Bar-Twenty Associates
17 N. Des Plaines Street
Chicago 6, IL
Licensed: Hopalong Cassidy Bank
Savings Plan
List of Licensees: 2/1/52, 5/19/54,
2/1/56, 6/20/56, 10/23/57, 4/10/58,
12/1/58

Brewster Leather Goods Corporation
101 West 31st Street
New York 1, NY
Licensed: Children's Western style
handled school bags
Date of Contract: 7/1/50
List of Licensees: 11/15/51, 2/1/52,
7/10/52

[Brewster Leather Goods Corporation]
117 East 24th Street
New York, NY
List of Licensees: Pocket Guide of
Hopalong Cassidy Western Wear No. 4
Spring 1953, 5/19/54, 2/1/56,
6/20/56, 10/23/57, 4/10/58

[Brewster Leather Goods Corporation]
440 Nepperhan Avenue
Yonkers, NY
List of Licensees: 12/1/58, 10/29/62

Brown & Bigelow
1286 University Avenue
St. Paul 4, MN
Licensed: Hopalong Cassidy and
Topper calendars
Date of Contract: 11/13/50
List of Licensees: 11/15/51, 7/10/52,
10/19/62

Buzza Cardozo Corporation
127 N. San Vincente Boulevard
Los Angeles 48, CA
Licensed: Greeting cards, party
invitations, notes, birth announcements
Date of Contract: 7/1/50
List of Licensees: 11/15/51, 2/1/52,
7/10/52, 5/19/54, 10/19/62

Capitol Records, Inc.
Sunset and Vine Streets
Hollywood 28, CA
Licensed: Record albums
List of Licensees: 2/1/52, Pocket Guide
of Hopalong Cassidy Western Wear No.
4 Spring 1953

**Casey Premium Merchandise
Company**
1132 South Wabash
Chicago, Il
Licensed: Pennants, banners, pillows
Date of Contract: 7/1/50
List of Licensees: 10/19/62

Philip Florin Company
55 Gouverneur Street
Newark 4, NJ
Licensed: Wallets, ring binders, spiral
bound notebooks, utility kits, fitted sets
Date of Contract: 7/1/50
List of Licensees: 8/1/50, 11/15/51,
2/1/52, 7/10/52, 5/19/54, 2/1/56,
6/20/56, 10/19/62

[Philip Florin, Inc.]
358 Fifth Avenue
New York 1, NY
List of Licensees: Pocket Guide of
Hopalong Cassidy Western Wear No. 4
Spring 1953

Hassenfeld Brothers, Inc.
Pawtucket, RI
Licensed: Scrap books, photo and
autograph albums, doctor's Western
play kit, plain and zippered pencil
pouches, pencils encased in wood
(except mechanical pencils), erasers,
rulers, pencil sharpeners, and school
children's pencil sets in paper, artificial
leather, leather, and plastic pencil boxes
Date of Contract: 9/1/52
List of Licensees: 11/15/51, 10/19/62

Lowe, Inc.
1324 Fifty-Second Street
Kenosha, WI
Licensed: Coloring books
Date of Contract: 8/1/50
List of Licensees: 11/15/51, 7/10/52,
10/19/62

The Parker Pen Company
Janesville, WI
Licensed: Hopalong Cassidy fountain
and/or ball point pens
Date of Contract: 8/15/50
List of Licensees: 11/15/51, 7/10/52,
10/19/62

Topper Toys, Inc.
2000 W. Fulton Street
Chicago, IL
Licensed: Home Savings Banks,
flashlights
Date of Contract: 7/1/50
List of Licensees: 11/15/51, 2/1/52,
10/19/62

**Western Printing and Lithographing
Company**
1220 Mound Avenue
Racine, WI
Licensed: Jigsaw puzzles,boxed Sticker
Fun cutout books and cutouts,
children's stationery
Date of Contract: 7/1/50
List of Licensees: 11/15/51, 2/1/52,
7/10/52, 4/10/58, 10/23/57, 10/19/62

TOILETRIES:

Associated Brands, Inc.
35 Claver Place
Brooklyn, NY
Licensed: Children's hair trainer lotion
Date of Contract: 5/1/51
List of Licensees: 11/15/51, 2/1/52,
7/10/52, Pocket Guide of Hopalong
Cassidy Western Wear No. 4 Spring
1953, 5/19/54, 2/1/56, 6/20/56,
10/23/57, 4/10/58

[Associated Brands]
50 Wallabout Street
Brooklyn 11, NY
List of Licensees: 12/1/58, 10/19/62

Daggett & Ramsdell, Inc.
420 Frelinghuysen Avenue
Newark, NJ
Licensed: Children's soap, bubble bath,
hair trainer, shampoo
Date of Contract: 7/1/50
List of Licensees: 11/15/51, 2/1/52,
7/10/52, Pocket Guide of Hopalong
Cassidy Western Wear No. 4 Spring
1953, 5/19/54, 2/1/56, 6/20/56,
10/23/57, 4/10/58, 12/1/58, 10/19/62

Rubicon, Inc
347 Fifth Avenue
New York 16, NY
Licensed: Combs, brushes, hair trainer,
spark shooter
List of Licensees: 11/15/51, 7/10/52

Weco Products Company
20 N. Wacker Drive
Chicago, IL
Licensed: Toothbrushes, toothpaste,
sets
List of Licensees: 11/15/51

TOYS AND NOVELTIES:

Aladdin Industries, Inc.
703 Murfreesboro Road
Nashville, TN
Licensed: Metal lunch box with
thermos, vacuum bottles
Date of Contract: 7/1/50
List of Licensees: 11/15/51, 2/1/52,
7/10/52, Pocket Guide of Hopalong
Cassidy Western Wear No. 4 Spring
1953, 5/19/54, 2/1/56, 6/20/56,
10/23/57, 4/10/58, 12/1/58, 10/19/62

All Metal Products Company
Wyandotte, MI
(New York Office: 200 Fifth Avenue)
Licensed: Single and double leather
holster sets, cuffs, metal toys including
cap pistols and spurs
Date of Contract: 7/1/50
List of Licensees: 11/15/51, 2/1/52,
7/10/52, Pocket Guide of Hopalong
Cassidy Western Wear No. 4 Spring
1953, 5/19/54, 2/1/56, 6/20/56,
10/19/62

American Toy & Furniture Company
6130 N. Clark Street
Chicago 26, IL
Licensed: Wood burning sets
Date of Contract: 7/1/50
List of Licensees: 11/15/51, 2/1/5,
Pocket Guide of Hopalong Cassidy
Western Wear No. 4 Spring 1953,
10/19/62

Ames Harris Neville Company
2800 Seventeenth Street
San Francisco 10, CA
Licensed: Children's tents and sleeping
bags
Date of Contract: 9/11/51
List of Licensees: 11/15/51, 2/1/52,
7/10/52, 10/19/62

**Arvin Industries, Inc. (formerly
Noblitt Sparks Industries, Inc.)**
Columbus, IN
Licensed: Hopalong Cassidy radios
Date of Contract: 1/28/50
List of Licensees: 11/15/51, 10/19/62

The H. J. Ashe Company, Inc.
37-66 Eighty-Second Street
Jackson Heights 72, NY
Licensed: Hopalong Cassidy flashlights
Date of Contract: 5/18/53
List of Licensees: 5/19/54, 2/1/56,
6/20/56, 10/23/57, 4/10/58, 12/1/58,

10/19/62

**Aubin-Lee Tex Rubber Products
Corporation of California**
13151 S. Western Avenue
Gardena, CA
Licensed: Inflatable rubber toys and
balloons
List of Licensees: 10/23/57, 4/10/58
See Also: *Lee-Tex Rubber Products
Corporation of California*

Automatic Toy Company
77 Alaska Street
Staten Island 10, NY
Licensed: Windup television toy,
shooting gallery game
Date of Contract: 7/1/50
List of Licensees: 11/15/51, 2/1/52,
10/19/62

Bayshore Industries, Inc.
Elkton, MD
Licensed: Caps for toy pistols
Date of Contract: 8/20/51
List of Licensees: 11/15/51, 2/1/52,
7/10/52, Pocket Guide of Hopalong
Cassidy Western Wear No. 4 Spring
1953, 10/19/62

Milton Bradley Company,
Springfield 2, MA
Licensed: Jigsaw puzzles, Buckaroo
game, inlay puzzles, regular children's
puzzles
Date of Contract: 7/1/50
List of Licensees: 11/15/51, 2/1/52,
7/10/52, Pocket Guide of Hopalong
Cassidy Western Wear No. 4 Spring
1953, 5/19/54, 2/1/56, 6/20/56, 10/19/62

B.W. Molded Plastics, Inc.
1346 East Walnut Street
Pasadena 4, CA
Licensed: Children's plastic toy blocks
Date of Contract: 6/15/51
List of Licensees: 11/15/51, 2/1/52,
7/10/5, Pocket Guide of Hopalong
Cassidy Western Wear No. 4 Spring
1953, 10/19/62

Charcook Company, Inc.
447 Los Feliz Boulevard
Glendale 4, CA
Licensed: Child's prefabricated Western
Ranch playhouse
Date of Contract: 9/12/50
List of Licensees: 10/19/62

Circle F Toys, Inc.
246 N. Central Avenue
Valley Stream, Long Island, NY
Licensed: Pogo sticks
Date of Contract: 8/23/51
List of Licensees: 11/15/51, 7/10/52

[Circle F Toys, Inc.]
111 Water Street
Brooklyn, NY
List of Licensees: 2/1/52, Pocket Guide
of Hopalong Cassidy Western Wear No.
4 Spring 1953, 5/19/54, 10/19/62

**Coltrin, McDonell & Corson/for Dyer
Products Company**
514 Second Street S.W.
Canton, OH
Licensed: Boys' home gym and muscle
builders
Date of Contract: 8/1/51
List of Licensees: 11/15/51, 7/10/52,
10/19/62
See Also: *Dyer Products Company*

Cowboy Juniors, Inc.
8554 Wilshire Boulevard

Beverly Hills, CA
Licensed: Children's Western trick rope
Date of Contract: 1/1/51
List of Licensees: 11/15/51, 2/1/52,
10/19/62

Charles W. Cradick Company
1169 N. Vermont Avenue
Los Angeles, CA
Licensed: Mechanical Topper rocking
horse
Date of Contract: 11/13/51
List of Licensees: 2/1/5, Pocket Guide
of Hopalong Cassidy Western Wear No.
4 Spring 1953, 10/19/62

Dyer Products Company
514 Second Street S.W.
Canton, OH
Licensed: Boys' home gym and muscle
builder
List of Licensees: 2/1/52
See Also: *Coltrin, McDonell & Corson*

Econolite Corporation
3517 W. Washington Blvd
Los Angeles, CA
Licensed: Rotating electric table lamp,
animated electric table lamp
Date of Contract: 7/1/50
List of Licensees: 11/15/51, 2/1/52,
Pocket Guide of Hopalong Cassidy
Western Wear No. 4 Spring 1953,
10/19/62

Einson-Freeman Company, Inc.
Starr & Bordon Avenues
Long Island City 1, NY
Licensed: Rubber band gun, blow toy,
paper mask, flip movies, 3-dimensional
cutout of Hoppy on Topper, paper
cowboy cuffs, Hoppy periscope, metal
badge, metal rings
List of Licensees: 11/15/51

**Joe Fredericksson Toy & Game
Manufacturing Company, Inc.**
3390 Seventeenth Street
San Francisco 10, CA
Licensed: Lithographed paper play
money
Date of Contract: 4/18/51
Termination Date: 12/31/52
List of Licensees: 11/15/51, 7/20/52,
10/19/62

**[Joe Fredericksson Toy & Game
Manufacturing Company, Inc.]**
831-833 Howard Street
San Francisco 3, CA
List of Licensees: 2/1/52

Galter Products Company
Herold Manufacturing Company,
Assignee
711 West Lake Street
Chicago, IL
Licensed: Cameras, binoculars
Date of Contract: 7/1/50
List of Licensees: 11/15/51, 2/1/52,
7/10/52, 10/19/62
See Also: *Herold Manufacturing
Company*

**D. P. Harris Hardware Manufacturing
Company**
99 Chambers Street
New York 7, NY
Licensed: Children's bicycles,
velocipedes, tribikes, sidewalk roller
skates, bicycle saddle bags
Date of Contract: 12/14/49
List of Licensees: 11/15/51, 2/1/52,
7/10/5, Pocket Guide of Hopalong
Cassidy Western Wear No. 4 Spring
1953, 5/19/54, 2/1/56, 6/20/56, 10/
23/57, 4/10/58, 12/1/58, 10/19/62

Herold Manufacturing Company
711 W. Lake Street
Chicago 6, IL
Licensed: Cameras, field glasses and/
or binoculars
List of Licensees: Pocket Guide of
Hopalong Cassidy Western Wear No. 4
Spring 1953, 5/19/54, 2/1/56,
6/20/56, 10/23/57
See Also: Galter Products Company

Ideal Novelty and Toy Company
200 Fifth Avenue
New York 10, NY
Licensed: Dolls styled as cowboys,
horses (some mechanical), plastic rifle
with metal parts
Date of Contract: 8/1/51
List of Licensees: 11/15/51, 2/1/52,
7/10/52, Pocket Guide of Hopalong
Cassidy Western Wear No. 4 Spring
1953, 5/19/54, 2/1/56, 6/20/56,
10/23/57, 4/10/58, 12/1/58, 10/19/62

Imperial Knife Company, Inc.
14 Blount Street
Providence, RI
Licensed: Boys' pocket knives, juvenile
stainless steel knife, fork and spoon
sets
Date of Contract: 7/1/50
List of Licensees: 11/15/51, 2/1/52,
7/10/52, Pocket Guide of Hopalong
Cassidy Western Wear No. 4 Spring
1953, 5/19/54, 2/1/56, 6/20/56, 10/
23/57, 4/10/58, 12/1/58, 10/19/62

Laurel Ann Arts
1307 Riverside Drive
Los Angeles 31, CA
Licensed: Plaster statuette, paint set for
statuette
List of Licensees: 11/15/51

Lee-Tex Rubber Products of California
321 Jackson Street
Los Angeles 12, CA
Licensed: Inflatable rubber toys and
balloons
Date of Contract: 2/16/51
List of Licensees: 11/15/51, 2/1/52,
7/10/5, Pocket Guide of Hopalong
Cassidy Western Wear No. 4 Spring
1953, 5/19/54
*See Also: Aubin-Lee Tex Rubber
Products of California*

**[Lee-Tex Rubber Products of
California]**
13151 S. Western Avenue
Gardena, CA
List of Licensees: 2/1/56, 6/20/56,
10/23/57, 4/10/58, 12/1/58, 10/19/62

**Lee-Tex Rubber Products Corporation
of Illinois**
2157 North Damen Avenue
Chicago, IL
Licensed: Inflatable rubber toys and
balloons
List of Licensees: 5/19/54, 2/1/56,
6/20/56, 10/23/56, 4/10/58

**[Lee-Tex Rubber Products Corporation
of Illinois]**
1711 Terra Cotta Place
Chicago 145, IL
List of Licensees: 12/1/58

Louis Marx and Company
200 Fifth Avenue
New York, NY
Licensed: Mechanical toys, games

National Mask & Puppet Corporation
594-596 Pacific Street
Brooklyn, NY
Licensed: Hand and string puppets, toy
chest hassock
Date of Contract: 7/1/50
List of Licensees: 11/15/51, 2/1/52,
7/10/52, Pocket Guide of Hopalong
Cassidy Western Wear No. 4 Spring
1953

**[National Mask & Puppet Corpora-
tion]**
62 Eighteenth Street
Brooklyn 32, NY
List of Licensees: 10/19/62

Pacific Playing Card Company
948 South Flower Street
Los Angeles, CA
Licensed: Hopalong Cassidy games,
Hopalong Canasta card game
Date of Contract: 7/1/50
List of Licensees: 11/15/51, 2/1/52,
10/19/62

David B. Perlin
2112 S. Michigan Avenue
Chicago, IL
Licensed: Bicycle horn simulating a
steer horn
Date of Contract: 7/1/50
List of Licensees: 11/15/51, 10/19/62

**R & S Toy Manufacturing Company,
Inc.**
19-21 Heyward Street
Brooklyn 11, NY
Licensed: Hopalong Cassidy holsters
and holster and gun sets
Date of Contract: 4/1/57
List of Licensees: 10/23/57, 4/10/58,
12/1/58, 10/19/62

Rich Industries, Inc.
Clinton, IA
Licensed: Rocking horse

Rubbertone Corporation
416 Wall Street
Los Angeles 13, CA
Licensed: drums

Sawyer's, Inc.
755 West 20th Street
Portland, OR
Licensed: Stereoscopic viewer reels
and/or cards
Date of Contract: 5/1/51
List of Licensees: 11/15/51, 2/1/52,
7/10/52

[Sawyer's, Inc.]
735 S.W. 20th Place
Portland 7, OR
List of Licensees: Pocket Guide of
Hopalong Cassidy Western Wear No. 4
Spring 1953

[Sawyer's, Inc./ Tru-Vue Company]
P.O. Box 490
Portland, OR
List of Licensees: 5/19/54, 2/1/56,
6/20/56, 10/23/57, 4/10/58, 12/1/58,
10/19/62
See Also: Tru-Vue Company

**George Schmidt Manufacturing
Company**
716 East 14th Street
Los Angeles, CA
Licensed: guns, spurs

Schoellkopf Company
Dallas 2, TX

Licensed: Children's saddles
Date of Contract: 7/1/50
List of Licensees: 11/15/51, 2/1/52,
7/10/5, Pocket Guide of Hopalong
Cassidy Western Wear No. 4 Spring
1953, 10/19/62

Southwester Company
2130-40 S. Kedzie Avenue
Chicago 23, IL
Licensed: Fishing rods, reels, and sets,
toy fishing rods
Date of Contract: 7/1/51
List of Licensees: 11/15/51, 2/1/52,
7/10/52, Pocket Guide of Hopalong
Cassidy Western Wear No. 4 Spring
1953, 10/19/62

Stephens Products Company, Inc.
Middletown, CT
Licensed: Auto-Magic toy simulating
metal gun, containing twelve 16mm
frames
Date of Contract: 7/1/50
List of Licensees: 11/15/51, 2/1/52,
7/10/52, 10/19/62

Tigrett Industries
21 W. Superior Street
Chicago 10, IL
Licensed: Toys

Toy Enterprises of America
4635 S. Harlem Avenue
Berwyn, IL
Licensed: Magne-Safe Arrow magnetic
dart game
Date of Contract: 7/15/53
List of Licensees: 11/15/51, 2/1/52,
7/10/52, Pocket Guide of Hopalong
Cassidy Western Wear No. 4 Spring
1953

[Toy Enterprises of America]
4448 Elston Avenue
Chicago, IL
List of Licensees: 5/19/54, 10/19/62

Transogram Company, Inc.
200 Fifth Avenue
New York 10, NY
Licensed: Crayon and paint sets, Ring-
toss game, horseshoe pitching sets,
beanbags, slates, chalk, toy pastry
cooking pans, television chairs
Date of Contract: 3/14/50
List of Licensees: 11/15/51, 2/1/52,
7/10/52, Pocket Guide of Hopalong
Cassidy Western Wear No. 4 Spring
1953, 5/19/54, 10/19/62

Traveler Trading Company
926 Broadway
New York, NY
Licensed: Rubber latex masks

Tru-Vue Company
P.O. Box 490
Portland, OR
Licensed: Stereoscopic viewer reels
and/or cards
See Also: Sawyer's Inc.

Tumbleweed Togs
128 Wheeler Street
Arcadia, CA
Licensed: Holsters, cuffs
List of Licensees: 8/1/50

United States Time Corporation
500 Fifth Avenue
New York 18, NY
Licensed: Children's pocket watches,
wristwatches, alarm clocks
Date of Contract: 7/1/50
List of Licensees: 11/15/51, 2/1/52,

7/10/52, Pocket Guide of Hopalong
Cassidy Western Wear No. 4 Spring
1953

[United States Time Corporation]
Waterbury, CT
List of Licensees: 5/19/54, 2/1/56,
6/10/56, 10/23/57, 4/10/58, 12/1/58,
10/19/62

Urb Plastics Corporation
130 Prince Street
New York 12, NY
Licensed: Bouncing type vinyl plastic
inflated Hobby Horse with automatic
noise maker and wading pool
Date of Contract: 7/2/51
List of Licensees: 11/15/51, 2/1/52,
7/10/52, Pocket Guide of Hopalong
Cassidy Western Wear No. 4 Spring
1953, 5/19/54, 2/1/56, 6/20/56,
4/10/58, 12/1/58, 10/19/62

U. S. Fiber & Plastics Corporation
Union Street
Stirling, NJ
Licensed: Inflatable vinyl plastic horse
flotation toy for bathtub or pool
Date of Contract: 7/1/50
List of Licensees: 10/19/62

AUSTRALIAN
LICENSEES

FOOD PRODUCTS

Bester's Sweets Pty., Ltd.
Ballarat Road
Braybrook, Victoria
Licensed: Hopalong Cassidy Bar 20
chocolate coated macaroon bar,
Hopalong Cassidy Ranch toffee in a
Hopalong Cassidy printed tin
List of Licensees: 8/1/55, 2/1/56,
6/20/56, 11/29/56

Fyna Foods Pty., Ltd.
40 Yarra Street
Abbotsford, Victoria
Licensed: Toffee lollipops, assorted
wrapped toffee (loose and boxed),
Hoppy-ade powdered fruit drink,
Hoppy Ring Stick
List of Licensees: 11/15/51, 7/10/52,
8/1/55, 2/1/56, 6/20/56, 11/29/56

Keith Harris & Company, Ltd.
51 William Street
Melbourne, Victoria
Licensed: Hoppy Cola
List of Licensees: 8/1/55, 2/1/56,
6/20/56, 11/29/56

MERCHANDISE

CLOTHING:

Casual Clothing Company Pty., Ltd.
69 Camberwell Road
Camberwell, Victoria
Licensed: Boys' cloth frontier suit
ensemble, boys' and girls' cloth
cowboy style shirts, girls' Western style
cloth blouse, skirt and bolero tops
List of Licensees: 11/15/51, 7/10/52,
8/1/55

Davies Coop Limited
Melbourne
Licensed: Knitted cotton fleece-lined
sweatshirts and outerwear, knitted

cotton, rayon and cotton and rayon mix underwear and nightshirts and T shirts with or without collar or front opening
List of Licensees: 11/15/51, 7/10/52

Goldie Modes
271 Lonsdale Street
Melbourne C. 1
Licensed: Boys' and girls' jeans, jodphurs and boxer shorts made of cotton denim
List of Licensees: 11/15/51, 7/10/52, 8/1/55

Simon Hickey & Sons Pty., Ltd.
12-20 Queen Street
Sydney
Licensed: Belts, brace neckwear (ties), kerchiefs, toggles (tie rings or conchos), cuffs and anklets, cowboy chaps and vest and cowgirl skirt and sleeveless bolero ensembles made of leather, suede and/or leatherette, wallets and/or billfolds
List of Licensees: 11/15/51, 7/10/52, 8/1/55, 2/1/56, 6/20/56, 11/29/56

Luton Mens Hats Pty., Ltd.
39 Liverpool Street
Sydney
Licensed: Boys' and girls' wool, wool felt composition or straw cowboy hats
List of Licensees: 11/15/51, 7/10/52, 8/1/55, 2/1/56

Paddle Brothers Pty., Ltd.
4 Reid Street
North Fitzroy, Victoria
Licensed: Hopalong Cassidy leather boots and slippers
List of Licensees: 8/1/55, 6/20/56, 11/29/56

Messrs. Ratten, Cowland & Hare
349 Canterbury Road, Surry Hills
Melbourne
Licensed: Boys' and girls' dressing gowns
Licensee List: 11/15/51, 7/10/52

George Stooke & Son Pty., Ltd.
277 Flinders Lane
Melbourne
Licensed: Boys' wool, wool felt composition or straw cowboy hats with high or flat crowns, girls' wool, wool felt composition or straw cowboy hats with rounded crown
List of Licensees: 11/15/51, 7/10/52, 8/1/55, 2/1/56, 6/20/56, 11/29/56

Sutex Limited
49-59 Coppin Street
Richmond E1
Melbourne
Licensed: Boys' and girls' jacquard knitted sweaters and cardigans, swimsuits
List of Licensees: 11/15/51, 7/10/52

DOMESTICS:

Supertex Industries Limited
216 Henderson Road
Alexandria, N.S.W.
Licensed: Bedspreads, floor rugs, curtains, cushion covers, seat covers, bath mats, terry hand, face and bath towels, and terry yardage (all made from chenille, terry, ruggle-tuft, ruggle-twist and/or colonial weave material processes)
List of Licensees: 8/1/55, 2/1/56, 6/20/56, 11/29/56

STATIONERY:

Jean Littlewood
205 Alma Road
Elsternwick, Victoria
Licensed: Autograph books, photograph albums, scrap book albums, swap card albums
List of Licensees: 8/1/55, 2/1/56, 6/20/56, 11/29/56

Personality Merchandising
125 Swanston Street
Melbourne
Licensed: Greeting cards
List of Licensees: 7/10/52

Valentine Publishing Company Pty., Ltd.
214 Queen Street
Melbourne, C. 1
Licensed: Greeting cards, autographed post cards
List of Licensees: 8/1/55, 2/1/56, 6/20/56, 11/29/56

Warwick Manufacturing Company
59 Westminster Street
Oakleigh, Victoria
Licensed: Hopalong Cassidy black and white leather school bags
List of Licensees: 8/1/55, 2/1/56, 6/20/56, 11/29/56

TOYS AND NOVELTIES:

John Brent & Company
101 Bridge Road
Richmond E. 1, Victoria
Melbourne
Licensed: Cap pistols, holster and gun sets, spurs, large plastic carbine cap-firing rifle (25" long), small plastic clicker rifle (18" long), Hoppy Junior model pistols, sheriff's badge, novelty cowboy boot shaped whistle, Western trick lariat
List of Licensees: 11/15/51, 7/10/52, 8/1/55, 2/1/56

Containers, Ltd.
265 Franklin Street
Melbourne C. 1
Licensed: Jigsaw puzzles
List of Licensees: 6/20/56, 11/29/56

R. Dorning, Dawn Canvas Goods
512 Swanston Street
Carlton, N. 3 Victoria
Licensed: Children's canvas play tents
List of Licensees: 11/15/51, 7/10/52, 8/1/55, 2/1/56, 6/20/56, 11/29/56

Freydis Manufacturing Company
106-116 A'Beckett Street,
Melbourne C. 1
Licensed: Plastic Hopalong Cassidy lunch bags
List of Licensees: 8/1/55, 2/1/56, 6/20/56, 11/29/56

Keith Harris & Company, Ltd.
51 William Street
Melbourne, Victoria
Licensed: Cake decorations including molded figures and letters made from plastic, sugar compounds and/or plaster compounds, embossed seal used for decorating cake frills and/or ornaments, paper serviettes
List of Licensees: 8/1/55, 2/1/56, 6/20/56, 11/29/56

Janda Amorces Company
4 South Audley Street
Abbotsford, Victoria
Licensed: Repeating caps for toy pistols and rifles
List of Licensees: 2/1/56, 6/20/56, 11/29/56

Janda Trading Company
24 Atkin Street
North Melbourne N. 1
Licensed: Celluloid covered clip-on badges, battery operated torch in shape of cowboy pistol
List of Licensees: 11/15/51, 7/10/52, 8/1/55, 2/1/56, 6/20/56, 11/29/56

Lincoln Industries, Ltd.
P.O. Box 18
Newmarket, Auckland S.E. 1
New Zealand
Licensed: Pistols, holster sets, spurs, sheriff's badges, cowboy cuffs, gaiters, triangular scarves, trouser belts, scarf slides (toggles), denim jeans, leather or leatherette chaps and vest ensembles
List of Licensees: 2/1/56, 6/20/56, 11/29/56

W. Owen Pty., Ltd.
9 Armstrong Street North
Ballarat, Victoria
Licensed: Hopalong Cassidy Game
List of Licensees: 8/1/55, 2/1/56, 6/20/56, 11/29/56

Alex Tolmer & Associates Pty., Ltd.
29 Queens Bridge Street
South Melbourne
Licensed: Cap pistols, spurs, holster gun sets, sheriff's badge, novelty whistle in shape of cowboy boot, Hoppy Junior model pistol, large plastic carbine cap-firing rifle, small plastic clicker-rifle
List of Licensees: 6/20/56, 11/29/56

CANADIAN LICENSEES

FOOD PRODUCTS

BAKERIES:

W. T. Lynch & Sons
Sydney, Nova Scotia
Territory: Nova Scotia city of Sydney and 150 mile radius
Date of Contract: 12/6/54
List of Licensees: 2/8/55, 3/15/55, 1/20/56, 6/20/56, 10/19/62

Walker's Bread Limited
Ottawa, Ontario
Territory: Ontario city of Ottawa and 125 mile radius
Date of Contract: 1/14/52
List of Licensees: 7/10/52, 2/8/55, 3/15/55, 10/19/62

DAIRIES:

Orillia Dairy Company
24 West Street N.
Orillia, Ontario
Territory: Ontario city of Orillia
Licensed: Milk
Date of Contract: 10/11/54
List of Licensees: 2/8/55, 3/15/55, 10/19/62

MISCELLANEOUS FOOD:

John Stuart Sales Company, Ltd.
7 Duke Street
Toronto, Ontario
Licensed: Frostade powdered soft drink concentrate
List of Licensees: 4/1/55, 5/19/54, 2/1/56

MERCHANDISE

CLOTHING:

Acme Glove Works, Ltd.
Montreal, Quebec
Licensed: Boys' and girls' leather Western gloves, boys' and girls' snowsuits
List of Licensees: 11/15/51

Adam Hats Limited
Long Branch, Ontario
Licensed: Boys' and girls' hats
List of Licensees: 11/15/51, 7/10/52, Pocket Guide of Hopalong Cassidy Western Wear No. 4 Spring 1953

Anson, Inc.
24 Baker Street
Providence 5, RI
Licensed: Boys' jewelry and accessories including chains, cuff links, money clips and tie holders
Date of Contract: 7/1/50
List of Licensees: 8/1/50, 11/15/51, 2/1/52, 7/10/52, 10/19/62

Barclay Knitwear Company
1239 Broadway
New York, NY
Licensed: Jacquards and sweaters including plain knit wool, and/or synthetic fibre, and/or cotton, long sleeve, short sleeve, and sleeveless
Date of Contract: 7/1/50
List of Licensees: 8/1/50, 11/15/51, 2/1/52, 7/10/52, 10/19/62

Belt Manufacturing Company of Canada, Ltd.
56 Esplanade
Toronto, Ontario
Licensed: Children's leather belts and suspenders, Western style strap handle school bags, wallets
List of Licensees: 11/15/52, 7/10/52, Pocket Guide of Hopalong Cassidy Western Wear No. 4 Spring 1953, 5/19/54, 2/1/56, 6/20/56, 11/29/56, 10/23/57, 4/10/58

Canada West Shoe Manufacturing Company, Ltd.
215 Watt Street
Winnipeg, Manitoba
Licensed: Children's Western leather boots
List of Licensees: 11/15/51, 7/10/52, Pocket Guide of Hopalong Cassidy Western Wear No. 4 Spring 1953

Fashion Hat and Cap Company
90 Chestnut Street
Toronto, Ontario
Licensed: Hats
List of Licensees: 5/19/54, 2/1/56, 6/20/56, 11/29/56

Joseph Gould & Sons, Ltd.
93 Spadina Avenue
Toronto, Ontario
Licensed: Boys' and girls' Western

shirts (excluding T shirts and polo shirts), boys' and girls' Western pants and jackets
List of Licensees: 11/15/51

Hat Stay Enterprises
Long Branch, Ontario
Licensed: Children's Western neckwear, ties and handkerchiefs
List of Licensees: 11/15/51

Herman Iskin & Company, Inc.
Souderton, PA
Licensed: Boys' and girls' cowboy and cowgirl playsuit ensembles consisting of collapsible cotton hat, pants or skirt, vest, shirt, kerchief, lariat, belt holster or pistol; chaps and vest ensemble and skirt and vest ensemble, boys' boxed set ensemble consisting of pair of boxer-type longees, pair of boxer-type shorts and shirt
List of Licensees: 8/1/50, 11/15/51, 7/10/52

Ray-Knit Manufacturing Company, Ltd.
680 King Street West
Toronto, Ontario
Licensed: Sweatshirts, T shirts, knitted pajamas, underwear
List of Licensees: 11/15/51, 7/10/52, 5/19/54, 2/1/56, 6/20/56, 11/29/56

Superior Converters, Ltd.
66 Portland Street
Toronto 28, Ontario
Licensed: Boys' and girls' dungarees, jeans, pedal pushers, and/or shorts
List of Licensees: 11/15/51, 7/10/52, Pocket Guide of Hopalong Cassidy Western Wear No. 4 Spring 1953, 11/29/56

[Superior Converters, Ltd.]
69 York Street
Toronto 1, Ontario
List of Licensees: 5/19/54, 2/1/56, 6/20/56

[Superior Converters, Ltd.]
310 Spadina Avenue
Toronto 2B, Ontario
List of Licensees: 10/23/57, 4/10/58

DOMESTICS:

Kosi-Karri-Krib Company
1654 O'Connor Drive
Toronto, Ontario
Licensed: Drapes, bedspreads, toy hassock, covering for youth bed headboard, raincoats
List of Licensees: 5/19/54

FURNITURE:

Morgan Manufacturing Company
Ashville, NC
Licensed: Children's bedroom furniture including beds, dressers, clothes closets, vanity cases, chests of drawers, combination desk/chest of drawers, chairs, horse rocker chairs, regular chairs and tables
Date of Contract: 7/1/50
List of Licensees: 11/15/51, 7/10/52

TOYS AND NOVELTIES:

Aladdin Industries
1401 The Queensway
Toronto, Ontario
Licensed: Lunch kits
List of Licensees: Pocket Guide of

Hopalong Cassidy Western Wear No. 4 Spring 1953

Imperial Knife Company, Inc.
14 Blount Street
Providence, RI
Licensed: Boys' pocket knives, juvenile stainless steel knife, fork and spoon sets
Date of Contract: 7/1/50
List of Licensees: 11/15/51, 2/1/52, 7/10/52, 5/19/54, 2/1/56, 6/20/56, 10/23/57, 4/10/58, 12/1/58, 10/19/62

Jeanne Figurine Company
168 Main Street
Toronto 13, Ontario
Licensed: Rubber molding kit
List of Licensees: 11/15/51, 7/10/52

Londontoys
1161 King Street
London, Ontario
Licensed: Western guns, spurs, cuffs and holster sets
List of Licensees: 11/15/51, 7/10/52, Pocket Guide of Hopalong Cassidy Western Wear No. 4 Spring 1953

Reliable Toy Company, Ltd.
258 Carlaw Avenue
Toronto 8, Ontario
Licensed: Stuffed dolls, plastic horse and rider toy
List of Licensees: 11/15/51, 7/10/52

Timex of Canada
1300 Jane Street
Toronto 9, Ontario
Licensed: Watches, toy watches, alarm clocks
List of Licensees: 10/23/57, 4/10/58

CUBAN LICENSEES

CLOTHING:

Barclay Knitwear Company
1239 Broadway
New York, NY
Licensed: Jacquards and sweaters including plain knit wool, and/or synthetic fibre, and/or cotton, long sleeve, short sleeve, and sleeveless
Date of Contract: 7/1/50
List of Licensees: 8/1/50, 11/15/51, 2/1/52, 7/10/52, 10/19/62

ENGLISH LICENSEES

FOOD PRODUCTS

Barratt & Company, Ltd.
Mayes Road, Wood Green
London N. 22
Licensed: Confectioneries
List of Licensees: The Toy Trader & Exporter 5/57

Lyon Manufacturing & Trading Company, Ltd.
Kingsway Corner Buildings
109 Kingsway
London W.C. 2
Licensed: Chocolate gold and silver coins, alphabet letters, dominoes, chess pieces, lollipops, figures (hollow, solid or filled), jigsaw puzzles and Easter eggs, sugar Easter eggs, sugar panorama eggs or figures
List of Licensees: 7/10/52, 5/19/54, 2/1/56

[Lyon Manufacturing & Trading Company, Ltd.]
79 Bow Road
London E. 3
List of Licensees: 6/20/56, 11/29/56

MERCHANDISE

CLOTHING:

The Cleveleys Shoe & Slipper Company, Ltd.
Dorset Avenue, Thornton Gt.
Cleveleys, Lancashire
Licensed: Bedroom slippers with plastic spurs
List of Licensees: The Toy Trader & Exporter 5/57

D. Dekker
167 Caledonian Road
London, N. 1
Licensed: Felt hats, ensembles consisting of trousers, skirts, boleros, shirts, kerchiefs, toggles, and/or cuffs, Hopalong Cassidy Tent Bunk House
List of Licensees: 6/20/56, The Toy Trader & Exporter, 5/57

Guisborough Shirt Company, Ltd.
West End
Guisborough, Yorkshire
Licensed: Pajamas, shirts, frontier suits, shirts, blouses
List of Licensees: 7/10/52, 5/19/54

Hamilton Carhartt, Ltd.
Carolina Port, Dundee
Licensed: Boys' and girls' cotton denim jeans and/or dungarees (sold as single units only)
List of Licensees: 5/19/54

Healthguard
Woodbridge Road
Leicester
Licensed: Underwear, sweater, socks, stole
List of Licensees: 5/19/54

Helm Knitwear
40 Crafton Street
Leicester
Licensed: Knitted pajamas, knitted T shirts, knitted and woven ties and scarves
List of Licensees: 7/10/52

Messrs. Sacul Hat Company, Ltd.
322 High Street North
East Ham, London
Licensed: Felt hats, playsuit ensembles
List of Licensees: 11/15/51

[Sacul Playsuits, Ltd.]
64-66 Sprowston Mews, Norwich Road
Forest Gate, London E. 7
List of Licensees: 7/10/52, 2/1/56, 6/20/56

[Sacul Playsuits, Ltd.]
5 Melson Street
Luton, Bedfordshire
List of Licensees: 5/19/54

Howard Wall Limited
25/27 Hackney Road
London, E. 2
Licensed: Braces (suspenders), elastic and leather belts, tape measure
List of Licensees: 7/10/52, 5/19/54, 2/1/56, 6/20/56, 11/29/56, The Toy Trader & Exporter 5/57

S. W. Wells & Sons, Ltd.
City Road
London E.C. 1
Licensed: Children's ties
List of Licensees: 5/19/54, 2/1/56

DOMESTICS:

Trail Trading Company, Ltd.
69 Rivington Street
London E.C. 2
Licensed: Wool, cotton, fibre, and/or rayon rugs
List of Licensees: 5/19/54, 2/1/56, 6/20/56, 11/29/56

PROMOTIONAL:

Young & Rubicam, Ltd., Advertising
Agency for Mars Limited
Roxburghe House
285 Regent Street
London W. 1
Licensed: Point-of-sale material for Mars Spangles Candy Bars
List of Licensees: 5/19/54, 2/1/56, 6/20/56

STATIONERY:

Adprint Limited
68 Oxford Street
London W. 1
Licensed: Children's books
List of Licensees: The Toy Trader & Exporter 5/57

Amalgamated Press Ltd.
Fleetway House
Farringdon Street, E.C. 4
Licensed: Hopalong Cassidy strip stories in "Knockout"
List of Licensees: The Toy Trader & Exporter 5/57

Dailley & Company, Ltd.
Northumberland Park
London N. 17
Licensed: Paper party tableware including napkins, doilies, tablecloths, plates, cups, shelf paper, and crimped and pleated cake and jelly containers
List of Licensees: 7/10/52, 5/19/54, 2/1/56, 6/20/56, 11/29/56

Messrs. C. W. Faulkner & Company, Ltd.
28 City Road
London E.C. 1
Licensed: Calendars, greeting cards
List of Licensees: 7/10/52, 5/19/54, 2/1/56, 6/20/56

Hodder & Stoughton Limited
St. Paul's House
Warwick Square
London, E.C. 4
Licensed: Novels
List of Licensees: The Toy Trader & Exporter 5/57

L. Miller & Son Limited
344 Hackney Road
London, E. 2
Licensed: Monthly comic
List of Licensees: The Toy Trader & Exporter 5/57

Messrs. J. C. Rose (London) Ltd.
30 City Road
London E.C. 1
Licensed: Wooden pencils with black lead or colored slips, plain or decorated, assorted lengths
List of Licensees: 5/19/54, 2/1/56

Messrs. Valentine & Son, Ltd.
P.O. Box 74
Westfield Works
Dundee, Scotland
Licensed: Greeting cards, spiral notebook, printed lampshade (F. W. Woolworth & Co., Ltd., only)
List of Licensees: 6/20/56, 11/29/56, The Toy Trader & Exporter 5/57

Walsall Lithographic Company, Ltd.
Midland Road
Walsall
Licensed: Hopalong Cassidy color prints
List of Licensees: The Toy Trader & Exporter 5/57

Western Printing and Lithographing Company
1220 Mound Avenue
Racine, WI
Licensed: Jigsaw puzzles, Sticker Fun Book
List of Licensees: 2/1/56, 6/20/56, 11/29/56

TOYS AND NOVELTIES:

Adprint Limited
51 A Rathbone Place
London W. 1
Licensed: Cardboard frame tray puzzles
List of Licensees: 6/10/56, 11/29/56

The Chad Valley Company, Ltd.
Harborne
Birmingham 17.
Licensed: School slate outfit, crayon and stencil sets, Canasta, Chinese Checkers, target practice, picture dominoes, marionette, television set, Shooting Gallery, shooting games, Bagatelle
List of Licensees: 7/10/52, 5/19/54, 2/1/56, 6/20/56, 11/29/56

D. Dekker
167 Caledonian Road
London, N. 1
Licensed: Hopalong Cassidy Tent Bunk House
List of Licensees: 6/20/56, 11/29/56, The Toy Trader & Exporter 5/57

Die Casting Machine Tools, Ltd.
152 Green Lanes
London, N. 13
Licensed: Pistols, spurs with straps, pistols with plastic holster, lapel badges.
List of Licensees: 6/20/56, 11/29/56

Efroc Limited
55 The Mall
Ealing W. 5
London
Licensed: Wood or cardboard jigsaw puzzles
List of Licensees: 11/15/51, 7/10/52

Electrical & Musical Industries Limited (Capitol Records)
8-11 Great Castle Street
London, W. 1
Licensed: Hopalong Cassidy children's records
List of Licensees: The Toy Trader & Exporter 5/57

G. B. Equipments Limited
Film Library
1 Aintree Road
Perivale, Middlesex
Licensed: Hopalong Cassidy films for home showing and purchase
List of Licensees: The Toy Trader & Exporter 5/57

Lewis Knight Limited
8 Chingford Mount Road
London, E. 4
Licensed: Balloons
List of Licensees: The Toy Trader & Exporter 5/57

Keith Lowe (Engineers)
68 King Street
Dudley, Worcester
Licensed: Die cast pistols and cap pistols, non-air type guns and rifles, pop guns
List of Licensees: 11/15/51, 2/1/56

[Keith Lowe Engineering]
Eclipse Works
Church Street
Dudley, Worcester
List of Licensees: 5/19/54

Model Toys, Ltd.
11 Coldborne Road
London, W. 10
Licensed: Toy cowboys, Indians, horses and ranch houses made of hollow-cast or solid metal, plastic or liquid wood
List of Licensees: 5/19/54, 2/1/56, 6/20/56, 11/29/56, The Toy Trader & Exporter 5/57

Messrs. George Sampson & Son, Ltd.
Liversedge Leather Works
Liversedge, Yorkshire
Licensed: Leather, leathercloth or plastic cuffs and wallets, leather holster sets, straps for saddles, spurs and roller skates, leather belts
List of Licensees: 11/15/51, 7/10/52, 5/19/54, 2/1/56, 6/20/56, 11/29/56, The Toy Trader & Exporter 5/57

Messrs. Seamer Products (Sculptorcraft) Ltd.
23-27 Eastbourne Street
Hull
Licensed: Boxed modeling set consisting of rubber molds for casting figures, individual Hopalong Cassidy statuettes made of materials other than metal
List of Licensees: 6/10/56, 11/29/56

Roy Stewart & Company, Ltd.
Buckingham House
19/21 Palace Street
Westminster, London S.W. 1
Licensed: Balloons with Hopalong Cassidy imprint (sizes 7 and 8)
List of Licensees: 6/20/56, 11/29/56

Toy Importers Limited - Timpo Toys
26 Westbourne Grove
London W. 2
Licensed: Toy cowboys, Indians, horses and ranch houses made of hollow-cast or solid metal, plastic or liquid wood
List of Licensees: 11/15/51, 7/10/52

Uniray (U.K.) Limited
32 Old Church Street
Chelsea, London, S.W. 3
Licensed: Paint by number sets
List of Licensees: The Toy Trader & Exporter 5/57

The United States Time Corporation
U.K. Time Branch
Campdown Road
Dundee, Scotland
Licensed: All types watches including wristwatches and pocket watches, alarm clocks
List of Licensees: 5/19/54, 2/1/56, 6/20/56, 11/29/56

United States Time Corporation - Timex
161/167 Oxford Street
London, W. 1
Licensed: Hopalong Cassidy "Timex" watches
List of Licensees: The Toy Trader & Exporter 5/57

Wells Brimtoy Distributors Limited
Progress Works
Stirling Road
Walthamstow, E. 17
Licensed: Shooting Game, musical hobby-horse
List of Licensees: The Toy Trader & Exporter 5/57

Young & Fogg Rubber Company, Ltd.
124 Haydons Road
London S.W. 19
Licensed: Rubber guns and balloons, hand puppets, plaster busts and plaques
List of Licensees: 7/10/52, 5/19/54, 2/1/56

NORWAY AND SWEDEN LICENSEES

CLOTHING:

AB Libo-Konfektion
Boras, Sweden
Licensed: Jeans, jacket, skirt, shirt, knitted sportswear
List of Licensees: 2/1/56, 6/20/56, 11/29/56

Gaula Fabrikker
Trondheim
Norway
Licensed: Children's cowboy pants
List of Licensees: 6/20/56, 11/29/56

Stinson A/B
Sandskrona, Sweden
Licensed: Leather belts
List of Licensees: 5/19/54, 2/1/56, 6/20/56, 11/29/56

TOY AND NOVELTIES:

AB Alga
Norrtullsgatan 63
Stockholm, Sweden
Licensed: Parlor games in Swedish
List of Licensees: 5/19/54, 2/1/56, 6/20/56, 11/29/56

Firma Imargo
Ringvagen 131
Stockholm So, Sweden
Licensed: Plastic bag, leather comb case, leather pen case with set of fountain pen, ballpoint pen and mechanical pencil, sheath knife with leather sheaf, leather wallet
List of Licensees: 5/19/54, 2/1/56, 6/20/56, 11/29/56

Mittet & Company A/S
Oslo, Norway
Licensed: Hopalong Cassidy Game
List of Licensees: 5/19/54, 2/1/56, 6/20/56, 11/29/56

SOUTH AMERICAN LICENSEES

FOOD PRODUCTS

Topps Chewing Gum, Inc.
237 Thirty-Seventh Street
Brooklyn 32, NY
Licensed: Chewing gum, hard candy, caramels
List of Licensees: 2/1/56

U.S. TERRITORIES & POSSESSIONS

CLOTHING:

Barclay Knitwear Company
1239 Broadway
New York, NY
Licensed: Jacquards and sweaters including plain knit wool, and/or synthetic fibre, and/or cotton, long sleeve, short sleeve, and sleeveless
Date of Contract: 7/1/50
List of Licensees: 8/1/50, 11/15/51, 2/1/52, 7/10/52, 10/19/62

DOMESTICS:

Paulsboro Manufacturing Company
Architects Building
17th and Sansom Streets
Philadelphia 3, PA
Licensed: Linoleum floor coverings and wainscoting
Date of Contract: 12/18/50
List of Licensees: 11/15/51, 2/1/52, 7/10/52, 5/19/54, 10/19/62

WORLD-WIDE LICENSEES

STATIONERY:

Western Printing and Lithographing Company
1220 Mound Avenue
Racine, WI
Licensed: Jigsaw puzzles, boxed Sticker Fun cutout books and cutouts, children's stationery
Date of Contract: 7/1/50
List of Licensees: 11/15/51, 2/1/52, 7/10/52, 4/10/58, 10/23/57, 10/19/62

TOYS AND NOVELTIES:

Sawyer's, Inc.
755 West 20th Street
Portland, OR
Licensed: Stereoscopic viewer reels and/or cards
Date of Contract: 5/1/51
List of Licensees: 11/15/51, 2/1/52, 7/10/52

FILMOGRAPHY

THE ORPHAN

Director:	J. Gordan Edwards
Producer:	William Fox
Associate Producer:	None credited
Photographer:	John W. Boyle
Film Editor:	None credited
Assistant Director:	None credited
Screenplay:	Roy Somerville
Source Novel:	Clarence E. Mulford, *The Orphan* (1908)
Distributor:	Fox
Copyright:	April 25, 1920
Length:	6 reels, 5695 feet

Cast

The Orphan:	William Farnum
Helen Shields:	Louise Lovely
Margaret Shields:	Olive White
Sheriff Jim Shields:	George Nicholls
Tex Willard:	Henry J. Hebert
Bucknell:	Earl Crain
Bill Howland:	G. Raymond Nye
Joe Sneed:	Harry DeVere
Martin:	Al Fremont
Aunt Cynthia:	Carrie Clark Ward

THE DEADWOOD COACH

Director:	Lynn Reynolds
Producer:	William Fox
Associate Producer:	None credited
Photographer:	Dan Clark
Film Editor:	None credited
Assistant Director:	None credited
Screenplay:	Lynn Reynolds
Source Novel:	Clarence E. Mulford, *The Orphan* (1908)
Distributor:	Fox
Copyright:	December 6, 1924
Length:	7 reels, 6346 feet

Cast

The Orphan:	Tom Mix
Tex Wilson:	George Bancroft
Jim Shields:	DeWitt Jennings
Helen Shields:	Doris May
Mrs. Shields:	Norma Wills
Matilda Shields:	Nora Cecil
Bill Howland:	Buster Gardner
Charlie Winter:	Lucien Littlefield
Sneed:	Sid Jordan
Walter Gordon:	Frank Coffyn
Mrs. Gordon:	Jane Keckley
Jimmie Gordon:	Ernest Butterworth

HOP-A-LONG CASSIDY

(aka Hopalong Cassidy Enters)

Director:	Howard Bretherton
Producer:	Harry Sherman
Associate Producer:	George Green
Photographer:	Archie Stout
Film Editor:	Edward Schroeder
Assistant Director:	None credited
Screenplay:	Doris Schroeder
Source Novel:	Clarence E. Mulford, *Hopalong Cassidy* (1910)
Distributor:	Paramount
Copyright:	August 20, 1935
Released:	July 30, 1935
Length:	60-63 minutes

Cast

Hopalong Cassidy:	William Boyd
Johnny Nelson:	Jimmy Ellison
Mary Meeker:	Paula Stone
Jim Meeker:	Robert Warwick
Uncle Ben:	George Hayes
Pecos Jack Anthony:	Kenneth Thomson
Red Conners:	Frank McGlynn, Jr.
Buck Peters:	Charles Middleton
Salem:	Willie Fung
Frisco:	Frank Campeau
Tom Shaw:	James Mason
Hall:	Ted Adams
Doc Riley:	Franklyn Farnum
Party Guest:	John Merton
Uncredited Bit Part:	Wally West

THE EAGLE'S BROOD

Director:	Howard Bretherton
Producer:	Harry Sherman
Associate Producer:	George Green
Photographer:	Archie Stout
Film Editor:	Edward Schroeder
Assistant Director:	Ray Flynn
Screenplay:	Doris Schroeder and Harrison Jacobs
Source Novel:	Clarence E. Mulford, *Hopalong Cassidy and the Eagle's Brood* (1931)
Distributor:	Paramount
Copyright:	October 25, 1935
Released:	October 10, 1935
Length:	59-65 minutes

Cast

Hopalong Cassidy:	William Boyd
Johnny Nelson:	Jimmy Ellison
El Toro:	William Farnum
Spike:	George Hayes
Big Henry:	Addison Richards
Dolores:	Nana Martinez (Joan Woodbury)
Mike:	Frank Shannon
Dolly:	Dorothy Revier
Steve:	Paul Fix
Pop:	Al Lyndell
Ed:	John Merton
Pablo:	George Mari
Esteban:	Juan Torena
Sheriff:	Henry Sylvester

BAR 20 RIDES AGAIN

Director:	Howard Bretherton
Producer:	Harry Sherman
Associate Producer:	George Green
Photographer:	Archie Stout
Film Editor:	Edward Schroeder
Assistant Director:	Ray Flynn
Screenplay:	Doris Schroeder and Gerald Geraghty
Source Novel:	Clarence E. Mulford, *The Bar-20 Rides Again* (1926)
Distributor:	Paramount
Copyright:	December 20, 1935
Released:	November 20, 1935
Length:	62-65 minutes

Cast

Hopalong Cassidy:	William Boyd
Johnny Nelson:	Jimmy Ellison
Margaret Arnold:	Jean Rouverol
Windy:	George Hayes
George Perdue/Nevada:	Harry Worth
Red Conners:	Frank McGlynn, Jr.
Jim Arnold:	Howard Lang
Clarissa Peters:	Ethel Wales
Buck Peters:	J. P. McGowan
Gila:	Paul Fix
Herb Layton:	Joe Rickson
Cinco:	Al St. John
Carp:	John Merton
Elbows:	Frank Layton
Bar 20 Hands:	Chill Wills and His Avalon Boys

CALL OF THE PRAIRIE

Director:	Howard Bretherton
Producer:	Harry Sherman
Associate Producer:	George Green
Photographer:	Archie Stout
Film Editor:	Edward Schroeder
Assistant Director:	None credited
Screenplay:	Doris Schroeder and Vernon Smith
Source Novel:	Clarence E. Mulford, *Hopalong Cassidy's Protege* (1926)
Distributor:	Paramount
Copyright:	March 6, 1936
Released:	March 6, 1936
Length:	65 minutes

Cast

Hopalong Cassidy:	William Boyd
Johnny Nelson:	Jimmy Ellison
Linda McHenry:	Muriel Evans
Shanghai McHenry:	George Hayes
Sheriff Sandy McQueen:	Chester Conklin
Sam Porter:	Al Bridge
Buck Peters:	Howard Lang
Wong:	Willie Fung
Tom:	Hank Mann

Slade:	Al Hill
Hoskins:	James Mason
Arizona:	John Merton
Bar 20 Hands:	Chill Wills and His Avalon Boys

THREE ON THE TRAIL

Director:	Howard Bretherton
Producer:	Harry Sherman
Associate Producer:	George Green
Photographer:	Archie Stout
Film Editor:	Edward Schroeder
Assistant Director:	Theodore Joss, D. M. (Derwin) Abrahams
Screenplay:	Doris Schroeder and Vernon Smith
Source Novel:	Clarence E. Mulford, *The Bar-20 Three* (1921)
Distributor:	Paramount
Copyright:	April 24, 1936
Released:	April 14, 1936
Length:	66-67 minutes

Cast

Hopalong Cassidy:	William Boyd
Johnny Nelson:	Jimmy Ellison
Pecos Kane:	Onslow Stevens
Mary Stevens:	Muriel Evans
Windy:	George Hayes
J. P. Ridley:	Claude King
Buck Peters:	William Duncan
Rose Peters:	Clara Kimball Young
Sheriff Sam Corwin:	John St. Polis
Idaho:	Ernie Adams
Kit Thorpe:	Al Hill
Jim Trask:	Ted Adams
Lewis:	John Rutherford
Conchita:	Lita Cortez
Rancher:	Hank Bell
Uncredited Bit Parts:	Lew Meehan, Artie Ortego, Franklyn Farnum

HEART OF THE WEST

Director:	Howard Bretherton
Producer:	Harry Sherman
Associate Producer:	George Green
Photographer:	Archie Stout
Film Editor:	Edward Schroeder
Assistant Director:	Theodore Joss, Phil Ford
Screenplay:	Doris Schroeder
Source Novel:	Clarence E. Mulford, *Mesquite Jenkins, Tumbleweed* (1932)
Distributor:	Paramount
Copyright:	July 10, 1936
Released:	July 24, 1936
Length:	60-63 minutes

Cast

Hopalong Cassidy:	William Boyd

Johnny Nelson:	Jimmy Ellison
Windy:	George Hayes
Big John Trumball:	Sidney Blackmer
Sally Jordan:	Lynn Gabriel
Jim Jordan:	Charles Martin
Barton:	Fred Kohler
Johnson:	Warner Richmond
Tom Paterson:	John Rutherford
Whitey:	Walter Miller
Saxon:	Ted Adams
Tim Grady:	Robert McKenzie
Judge:	Jim Elliott

HOPALONG CASSIDY RETURNS

Director:	Nate Watt
Producer:	Harry Sherman
Associate Producer:	Eugene Strong
Photographer:	Archie Stout
Film Editor:	Robert Warwick
Assistant Director:	V. O. Smith, M. (Derwin) Abrahams
Screenplay:	Harrison Jacobs
Source Novel:	Clarence E. Mulford, *Hopalong Cassidy Returns* (1924)
Distributor:	Paramount
Copyright:	October 16, 1936
Released:	October 12, 1936
Length:	71-75 minutes

Cast

Hopalong Cassidy:	William Boyd
Windy Halliday:	George Hayes
Mary Saunders:	Gail Sheridan
Robert Saunders:	John Beck
Lilli Marsh:	Evelyn Brent
Blackie Felton:	Stephen Morris (Morris Ankrum)
Buddy Cassidy:	William Janey
Peg Leg Holden:	Irving Bacon
Bob Claiborne:	Grant Richards
Luke:	Al St. John
Benson:	Ernie Adams
Buck:	Joe Rickson
Davis:	Ray Whitley
Dugan:	Claude Smith

TRAIL DUST

Director:	Nate Watt
Producer:	Harry Sherman
Associate Producer:	Eugene Strong
Photographer:	Archie Stout
Film Editor:	Robert Warwick
Assistant Director:	Harry Knight, D. M. (Derwin) Abrahams
Screenplay:	Al Martin
Source Novel:	Clarence E. Mulford, *Trail Dust* (1934)
Distributor:	Paramount
Copyright:	December 11, 1936
Released:	December 19, 1936
Length:	70-77 minutes

Cast

Hopalong Cassidy:	William Boyd
Johnny Nelson:	Jimmy Ellison
Windy Halliday:	George Hayes
Tex Anderson:	Stephen Morris (Morris Ankrum)
Beth Clark:	Gwynee Shipman
Lanky:	Britt Wood
Waggoner:	Dick Dickson
Red:	Earl Askam
Babson:	Al Bridge
Hank:	John Beach
Joe Wilson:	Ted Adams
Skinny:	Tom Halligan
Borden:	Dan Wolkiem
Lewis:	Harold Daniels
George:	Emmett Daly
Al:	Al St. John
Bowman:	Kenneth Harlan
Saunders:	George Chesebro
Bob:	Robert Drew
John Clark:	John Elliott

BORDERLAND

Director:	Nate Watt
Producer:	Harry Sherman
Associate Producer:	Eugene Strong
Photographer:	Archie Stout
Film Editor:	Robert Warwick
Assistant Director:	Harry Knight, D. M. (Derwin) Abrahams
Screenplay:	Harrison Jacobs
Source Novel:	Clarence E. Mulford, *Bring Me His Ears* (1922)
Distributor:	Paramount
Copyright:	February 26, 1937
Released:	February 26, 1937
Length:	82 minutes

Cast

Hopalong Cassidy:	William Boyd
Johnny Nelson:	Jimmy Ellison
Windy Halliday:	George Hayes
Loco/The Fox:	Stephen Morris (Morris Ankrum)
Dandy Morgan:	Al Bridge
Ranger Bailey:	John Beach
Grace Rand:	Nora Lane
Molly Rand:	Charlotte Wyatt
Colonel Gonzales:	Trevor Bardette
Major Stafford:	Earle Hodgins
Tom Parker:	George Chesebro
Doctor:	John St. Polis
Rancher:	Slim Whittaker
American Visitor:	Karl Hackett
American Visitor:	Robert Walker
Frank:	Frank Ellis
El Rio Sheriff:	J. P. McGowan
Gunman:	Cliff Parkinson
Gunman:	Edward Cassidy
Gunman:	Jack Evans

HILLS OF OLD WYOMING

Director:	Nate Watt
Producer:	Harry Sherman
Associate Producer:	None credited
Photographer:	Archie Stout
Film Editor:	Robert Warwick
Assistant Director:	D. M. (Derwin) Abrahams
Screenplay:	Maurice Geraghty
Source Novel:	Clarence E. Mulford, *The Round-Up* (1933)
Distributor:	Paramount
Copyright:	April 16, 1937
Released:	April 16, 1937
Length:	75-79 minutes

Cast

Hopalong Cassidy:	William Boyd
Windy Halliday:	George Hayes
Lucky Jenkins:	Russell Hayden
Andrews:	Stephen Morris (Morris Ankrum)
Alice Hutchins:	Gail Sheridan
Ma Hutchins:	Clara Kimball Young
Saunders:	John Beach
Thompson:	Earle Hodgins
Lone Eagle:	Steve Clemente
Chief Big Tree:	Chief John Big Tree
Peterson:	George Chesebro
Daniels:	Paul Gustine
Steve:	Leo McMahon
Smiley:	John Powers
Deputy:	James Mason

NORTH OF THE RIO GRANDE

Director:	Nate Watt
Producer:	Harry Sherman
Associate Producer:	None credited
Photographer:	Russell Harlan
Film Editor:	Robert Warwick
Assistant Director:	D. M. (Derwin) Abrahams, V. O. Smith
Screenplay:	Jack O'Donnell
Source Novel:	Clarence E. Mulford, *Cottonwood Gulch* (1924)
Distributor:	Paramount
Copyright:	June 18, 1937
Released:	June 18, 1937
Length:	65-70 minutes

Cast

Hopalong Cassidy:	William Boyd
Windy Halliday:	George Hayes
Lucky Jenkins:	Russell Hayden/ Henry Stoneham
The Lone Wolf:	Stephen Morris (Morris Ankrum)
Faro Annie:	Bernadene Hayes
Ace Chowder:	John Rutherford
Mary Cassidy:	Lorraine Randall
Bull O'Hara:	Walter Long
Mr. Wooden:	Lee Colt (Lee J. Cobb)
Deputy Jim Plunkett:	Al Ferguson
Clark:	John Beach
Joe:	Lafe McKee

RUSTLERS' VALLEY

Director:	Nate Watt
Producer:	Harry Sherman
Associate Producer:	None credited
Photographer:	Russell Harlan
Film Editor:	Sherman A. Rose
Supervising Film Editor:	Robert Warwick
Assistant Director:	D. M. (Derwin) Abrahams, V. O. Smith
Screenplay:	Harry O. Hoyt
Source Novel:	Clarence E. Mulford, *Rustlers' Valley* (1924)
Distributor:	Paramount
Copyright:	July 23, 1937
Released:	July 23, 1937
Length:	58-60 minutes

Cast

Hopalong Cassidy:	William Boyd
Windy Halliday:	George Hayes
Lucky Jenkins:	Russell Hayden
Glenn Randall:	Stephen Morris (Morris Ankrum)
Agnes Randall:	Muriel Evans
Taggart:	Ted Adams
Cal Howard:	Lee Colt (Lee J. Cobb)
Joe:	Al Ferguson
Sheriff Boulton:	John Beach
Clem Crawford:	Oscar Apfel
Party Guest:	Horace B. Carpenter
Stuttering Man:	John Powers
Uncredited Bit Part:	Bernadene Hayes
Uncredited Bit Part:	John St. Polis

HOPALONG RIDES AGAIN

Director:	Lesley Selander
Producer:	Harry Sherman
Associate Producer:	None credited
Photographer:	Russell Harlan
Film Editor:	Robert Warwick
Assistant Director:	D. M. (Derwin) Abrahams, Theodore Joos
Screenplay:	Norman Houston
Source Novel:	Clarence E. Mulford, *Black Buttes* (1923)
Distributor:	Paramount
Copyright:	August 20, 1937
Released:	September 30, 1937
Length:	63-65 minutes

Cast

Hopalong Cassidy:	William Boyd
Windy Halliday:	George Hayes
Lucky Jenkins:	Russell Hayden
Nora Blake:	Nora Lane
Prof. Horace Hepburn:	Harry Worth
Laura Peters:	Lois Wilde
Artie Peters:	Billy King
Buck Peters:	William Duncan
Blackie:	John Rutherford
Keno:	Ernie Adams
Pete:	John Beach
Slim:	Black Jack Ward

Rustler:	Frank Ellis
Rustler:	Ben Corbett
Uncredited Bit Part:	Artie Ortego

TEXAS TRAIL

Director:	David Selman
Producer:	Harry Sherman
Associate Producer:	None credited
Photographer:	Russell Harlan
Film Editor:	Sherman A. Rose
Assistant Director:	D. M. (Derwin) Abrahams, Theodore Joos
Screenplay:	Jack O'Donnell
Additional Dialogue:	Jack Merscreau
Source Novel:	Clarence E. Mulford, *Tex* (1922)
Distributor:	Paramount
Copyright:	November 26, 1937
Released:	November 26, 1937
Length:	58-60 minutes

Cast

Hopalong Cassidy:	William Boyd
Windy Halliday:	George Hayes
Lucky Jenkins:	Russell Hayden
Barbara Allen:	Judith Allen
Black Jack Carson:	Alexander Cross
Boots McCreedy:	Billy King
Major McCreedy:	Karl Hackett
Hawks:	Bob Kortman
Shorty:	Jack Rockwell
Smokey:	John Beach
Brad:	Raphael Bennett
Jordan:	Philo McCullough
General:	Earle Hodgins
Stockade Guard:	Ben Corbett
Lieutenant:	John Judd
Courier:	Clyde Kinney
Corporal:	Leo McMahon

PARTNERS OF THE PLAINS

Director:	Lesley Selander
Producer:	Harry Sherman
Associate Producer:	None credited
Photographer:	Russell Harlan
Film Editor:	Robert Warwick
Assistant Director:	D. M. (Derwin) Abrahams, Theodore Joos
Screenplay:	Harrison Jacobs
Source Novel:	Clarence E. Mulford, *The Man from Bar-20* (1918)
Distributor:	Paramount
Copyright:	January 28, 1938
Released:	January 14, 1938
Length:	68-70 minutes

Cast

Hopalong Cassidy:	William Boyd
Lucky Jenkins:	Russell Hayden
Baldy Morton:	Harvey Clark
Lorna Drake:	Gwen Gaze
Aunt Martha:	Hilda Plowright
Ronald Harwood:	John Warburton
Scar Lewis:	Al Bridge
Doc Galer:	All Hill
Sheriff:	Earle Hodgins
Mr. Benson:	John Beach
Uncredited Bit Part:	Jim Corey

CASSIDY OF BAR 20

Director:	Lesley Selander
Producer:	Harry Sherman
Associate Producer:	None credited
Photographer:	Russell Harlan
Film Editor:	Sherman A. Rose
Assistant Director:	D. M. (Derwin) Abrahams, Theodore Joos
Screenplay:	Norman Houston
Source Novel:	Clarence E. Mulford, *Me an' Shorty* (1929)
Distributor:	Paramount
Copyright:	February 25, 1938
Released:	February 25, 1938
Length:	56-60 minutes

Cast

Hopalong Cassidy:	William Boyd
Lucky Jenkins:	Russell Hayden
Pappy:	Frank Darien
Nora Blake:	Nora Lane
Clay Allison:	Robert Fiske
Tom Dillion:	John Elliott
Mary Dillion:	Margaret Marquis
Ma Caffrey:	Gretrude Hoffman
Jeff Caffrey:	Carleton Young
Judge Belcher:	Gordon Hart
Sheriff Hawley:	Edward Cassidy
Uncredited Bit Part:	Jim Toney

HEART OF ARIZONA

Director:	Lesley Selander
Producer:	Harry Sherman
Associate Producer:	J. D. Trop
Photographer:	Russell Harlan
Film Editor:	Sherman A. Rose
Assistant Director:	D. M. (Derwin) Abrahams, Theodore Joos
Screenplay:	Norman Houston
Distributor:	Paramount
Copyright:	April 22, 1938
Released:	April 22, 1938
Length:	68 minutes

Cast

Hopalong Cassidy:	William Boyd
Windy Halliday:	George Hayes
Lucky Jenkins:	Russell Hayden
Buck Peters:	John Elliott
Artie Peters:	Billy King
Belle Starr:	Natalie Moorehead
Jacqueline Starr:	Dorothy Short
Dan Ringo:	Alden (Stephen) Chase
Sheriff Hawley:	John Beach
Trimmer Winkler:	Lee Chandler
Twister:	Leo McMahon
Uncredited Bit Part:	Lee Phelps
Uncredited Bit Part:	Robert McKenzie

BAR 20 JUSTICE

Director:	Lesley Selander
Producer:	Harry Sherman
Associate Producer:	J. D. Trop
Photographer:	Russell Harlan
Film Editor:	Robert Warwick
Assistant Director:	D. M. (Derwin) Abrahams, Theodore Joos
Screenplay:	Arnold Belgard
Additional Dialogue:	Harrison Jacobs
Distributor:	Paramount
Copyright:	June 24, 1938
Released:	June 24, 1938
Length:	63-70 minutes

Cast

Hopalong Cassidy:	William Boyd
Windy Halliday:	George Hayes
Lucky Jenkins:	Russell Hayden
Ann Dennis:	Gwen Gaze
Buck Peters:	William Duncan
Frazier:	Pat (J.) O'Brien
Slade:	Paul Sutton
Denny Dennis:	John Beach
Perkins:	Joseph De Stefani
Pierce:	Walter Long
Ross:	H. Bruce Mitchell
Uncredited Bit Part:	Frosty Royce
Uncredited Bit Part:	Jim Toney

PRIDE OF THE WEST

Director:	Lesley Selander
Producer:	Harry Sherman
Associate Producer:	None credited
Photographer:	Russell Harlan
Film Editor:	Sherman A. Rose
Music Director:	Boris Morros
Assistant Director:	D. M. (Derwin) Abrahams, Theodore Joos
Screenplay:	Nate Watt
Distributor:	Paramount
Copyright:	July 8, 1938

Released:	July 8, 1938
Length:	55-56 minutes

Cast

Hopalong Cassidy:	William Boyd
Windy Halliday:	George Hayes
Lucky Jenkins:	Russell Hayden
Sheriff Tom Martin:	Earle Hodgins
Mary Martin:	Charlotte Field
Dick Martin:	Billy King
Caldwell:	Kenneth Harlan
Saunders:	George Strange
Nixon:	James Craig
Detective:	H. Bruce Mitchell
Sing Loo:	Willie Fung
Townsman:	George Morrell
Uncredited Bit Part:	Earl Askam
Uncredited Bit Part:	Jim Toney
Uncredited Bit Part:	Horace B. Carpenter
Uncredited Bit Part:	Henry Otho

IN OLD MEXICO

Director:	Edward D. Venturini
Producer:	Harry Sherman
Associate Producer:	None credited
Photographer:	Russell Harlan
Film Editor:	Robert Warwick
Music Score:	Gregory Stone
Music Director:	Boris Morros
Assistant Director:	D. M. (Derwin) Abrahams, Theodore Joos
Screenplay:	Harrison Jacobs
Distributor:	Paramount
Copyright:	September 9, 1938
Released:	September 9, 1938
Length:	62-67 minutes

Cast

Hopalong Cassidy:	William Boyd
Windy Halliday:	George Hayes
Lucky Jenkins:	Russell Hayden
Burke:	George Strange
The Fox:	Paul Sutton
Don Carlos Gonzales:	Allan Garcia
Anita Gonzales:	Jane (Jan) Clayton
Colonel Gonzales:	Trevor Bardette
Janet Leeds:	Betty Amann
Eleana:	Anna Demetrio
Pancho:	Tony Roux
Uncredited Bit Part:	Fred Burns
Uncredited Bit Part:	Cliff Parkinson

THE FRONTIERSMEN

Director:	Lesley Selander
Producer:	Harry Sherman
Associate Producer:	None credited
Photographer:	Russell Harlan
Film Editor:	Sherman A. Rose
Music Director:	Boris Morros
Assistant Director:	D. M. (Derwin) Abrahams, Theodore Joos
Screenplay:	Norman Houston
Additional Dialogue:	Harrison Jacobs
Distributor:	Paramount
Copyright:	December 16, 1938
Released:	December 16, 1938
Length:	71-74 minutes

Cast

Hopalong Cassidy:	William Boyd
Windy Halliday:	George Hayes
Lucky Jenkins:	Russell Hayden
June Lake:	Evelyn Venable
Mayor Judson Thorpe/ Dan Rawley:	Charles A. Hughes
Buck Peters:	William Duncan
Amanda Peters:	Clara Kimball Young
Miss Snook:	Emily Fitzroy
Artie Peters:	Dickie Jones
Quirt:	John Beach
Buster Sutton:	Jack Ward
School Boys:	St. Brendan Boys Choir
Uncredited Bit Part:	Black Jack Ward
Uncredited Bit Part:	George Morrell

Uncredited Bit Part:	Jim Corey

SUNSET TRAIL

Director:	Lesley Selander
Producer:	Harry Sherman
Associate Producer:	None credited
Photographer:	Russell Harlan
Film Editor:	Robert Warwick
Music Director:	Ed Paul
Assistant Director:	D. M. (Derwin) Abrahams, Theodore Joos
Screenplay:	Norman Houston
Distributor:	Paramount
Copyright:	February 24, 1939
Released:	February 24, 1939
Length:	60-67 minutes

Cast

Hopalong Cassidy:	William Boyd
Lucky Jenkins:	Russell Hayden
Windy Halliday:	George Hayes
Ann Marsh:	Charlotte Wynters
Dorrie Marsh:	Jane (Jan) Clayton
John Marsh:	Kenneth Harlan
Monte Keller:	Robert Fiske
Steve Dorman:	Anthony Nace
Abigail Snodgress:	Kathryn Sheldon
E. Prescott Furbush:	Maurice Cass
Superintendent:	Alphonse Ethier
Bouncer:	Glenn Strange
Mary Rogers:	Claudia Smith
Stage Driver:	Jack Rockwell
Patrol Captain:	Tom London
Uncredited Bit Part:	Jim Toney
Uncredited Bit Part:	Fred Burns
Uncredited Bit Part:	Jerry Jerome
Uncredited Bit Part:	Frank Ellis
Uncredited Bit Part:	Jim Corey
Uncredited Bit Part:	Horace B. Carpenter

SILVER ON THE SAGE

Director:	Lesley Selander
Producer:	Harry Sherman
Associate Producer:	J. D. Trop
Photographer:	Russell Harlan
Film Editor:	Robert Warwick
Music Director:	Boris Morros
Assistant Director:	D. M. (Derwin) Abrahams, Theodore Joos
Screenplay:	Maurice Geraghty
Distributor:	Paramount
Copyright:	March 31, 1939
Released:	March 31, 1939
Length:	66-68 minutes

Cast

Hopalong Cassidy:	William Boyd
Lucky Jenkins:	Russell Hayden
Windy Halliday:	George Hayes
Barbara Hamilton:	Ruth Rogers
Earl Brennan/Dave Talbot:	Stanley Ridges
Tom Hamilton:	Frederick Burton
City Marshal:	Jack Rockwell
Ewing:	Roy Barcroft
Pierce:	Edward Cassidy
Lane:	Wen Wright (Sherry Tansey)
Bartender:	H. Bruce Mitchell
Uncredited Bit Part:	Jim Corey
Uncredited Bit Part:	Hank Bell
Uncredited Bit Part:	George Morrell
Uncredited Bit Part:	Buzz Barton
Uncredited Bit Part:	Frank O' Connor
Uncredited Bit Part:	Herman Hack
Uncredited Bit Part:	Dick Dickinson

THE RENEGADE TRAIL

Director:	Lesley Selander
Producer:	Harry Sherman
Associate Producer:	None credited
Photographer:	Russell Harlan
Film Editor:	Sherman A. Rose
Music Director:	Boris Morros
Assistant Director:	D. M. (Derwin) Abrahams, Theodore Joos

Screenplay:	John Rathmell
Additional Dialogue:	Harrison Jacobs
Distributor:	Paramount
Copyright:	August 18, 1939
Released:	July 25, 1939
Length:	57-61 minutes

Cast

Hopalong Cassidy:	William Boyd
Lucky Jenkins:	Russell Hayden
Windy Halliday:	George Hayes
Mary Joyce:	Charlotte Wynters
Joey Joyce:	Sonny Bupp
Bob "Smoky" Joslin:	Russell Hopton
Stiff Hat Bailey:	Roy Barcroft
Tex Traynor:	John Merton
Red:	Eddie Dean
Slim Baker:	Jack Rockwell
Haskins:	Bob Kortman
Musicians:	The King's Men (Ken Darby, Rad Robinson, Jon Dobson, Bud Linn)

RANGE WAR

Director:	Lesley Selander
Producer:	Harry Sherman
Associate Producer:	None credited
Photographer:	Russell Harlan
Film Editor:	Sherman A. Rose
Music Director:	Victor Young
Assistant Director:	D. M. (Derwin) Abrahams
Screenplay:	Sam Robins
Additional Dialogue:	Walter Roberts
Story:	Josef Montiague
Distributor:	Paramount
Copyright:	September 8, 1939
Released:	September 8, 1939
Length:	64-66 minutes

Cast

Hopalong Cassidy:	William Boyd
Lucky Jenkins:	Russell Hayden
Speedy McGinnis:	Britt Wood
Padre Jose:	Pedro de Cordoba
Buck Collins:	Williard Robertson
Jim Marlow:	Matt Moore
Ellen Marlow:	Betty Moran
Charles Higgins:	Kenneth Harlan
Dave Morgan:	Francis McDonald
Pete:	Eddie Dean
Deputy:	Earle Hodgins
Rancher:	Jason Robards, Sr.
Agitator:	Stanley Price
Staley:	Raphael Bennett
Sheriff:	Glenn Strange
Felipe:	Don Latorre
Uncredited Bit Part:	George Chesebro

LAW OF THE PAMPAS

Director:	Nate Watt
Producer:	Harry Sherman
Associate Producer:	Joseph W. Engel
Photographer:	Russell Harlan
Film Editor:	Carrol Lewis
Music Director:	Victor Young
Assistant Director:	D. M. (Derwin) Abrahams
Screenplay:	Harrison Jacobs
Distributor:	Paramount
Copyright:	November 3, 1939
Released:	November 3, 1939
Length:	72-74 minutes

Cast

Hopalong Cassidy:	William Boyd
Lucky Jenkins:	Russell Hayden
Fernando Ramirez:	Sidney Toler
Chiquita:	Steffi Duna
Ralph Merritt:	Sidney Blackmer
Don Jose Valdez:	Pedro de Cordoba
Ernesto Valdez:	Jojo La Sadio
Buck Peters:	William Duncan
Dolores Ramirez:	Anna Demetrio
Curly Naples:	Eddie Dean

Slim Schultz:	Glenn Strange
Gaucho:	Tony Roux
Bolo Carrier:	Martin Garralaga
Musicians:	The King's Men

SANTE FE MARSHAL

Director:	Lesley Selander
Producer:	Harry Sherman
Associate Producer:	Joseph W. Engel
Photographer:	Russell Harlan
Film Editor:	Sherman A. Rose
Music Score:	John Leipold
Assistant Director:	D. M. (Derwin) Abrahams
Screenplay:	Harrison Jacobs
Distributor:	Paramount
Copyright:	January 26, 1940
Released:	January 26, 1940
Length:	65-68 minutes

Cast

Hopalong Cassidy:	William Boyd
Lucky Jenkins:	Russell Hayden
Ma Burton:	Marjorie Rambeau
Paula Bates:	Bernadene Hayes
Doc Bates:	Earle Hodgins
Axel:	Britt Wood
Blake:	Kenneth Harlan
Flint:	William Pagan
Tex Barnes:	George Anderson
John Gardner:	Jack Rockwell
Town Marshal:	Eddie Dean
Uncredited Bit Part:	Fred Graham
Uncredited Bit Part:	Matt Moore
Uncredited Bit Part:	Duke Green
Uncredited Bit Part:	Billy Jones
Uncredited Bit Part:	Tex Phelps
Uncredited Bit Part:	Cliff Parkinson

THE SHOWDOWN

Director:	Howard Bretherton
Producer:	Harry Sherman
Associate Producer:	None credited
Photographer:	Russell Harlan
Film Editor:	Carrol Lewis
Music Score:	John Leipold
Assistant Director:	D. M. (Derwin) Abrahams
Screenplay:	Harold and Daniel Kusell
Story:	Jack Jungmeyer
Distributor:	Paramount
Copyright:	March 8, 1940
Released:	March 8, 1940
Length:	63-65 minutes

Cast

Hopalong Cassidy:	William Boyd
Lucky Jenkins:	Russell Hayden
Speedy McGinnis:	Britt Wood
Baron Rendor:	Morris Ankrum
Sue Williard:	Jane (Jan) Clayton
Col. Rufus White:	Wright Kramer
Harry Cole:	Donald Kirke
Bowman:	Roy Barcroft
Marshal:	Eddie Dean
Johnson:	Kermit Maynard
Snell:	Walter Shumway
Musicians:	The King's Men

HIDDEN GOLD

Director:	Lesley Selander
Producer:	Harry Sherman
Associate Producer:	Joseph W. Engel
Photographer:	Russell Harlan
Film Editor:	Carrol Lewis
Music Director:	Irvin Talbot
Assistant Director:	D. M. (Derwin) Abrahams
Screenplay:	Gerald Geraghty, Jack Mersereau
Distributor:	Paramount
Copyright:	June 7, 1940
Released:	June 7, 1940
Length:	60-61 minutes

Cast

Hopalong Cassidy:	William Boyd
Lucky Jenkins:	Russell Hayden
Speedy McGinnis:	Britt Wood
Ed Colby:	Minor Watson
Jane Colby:	Ruth Rogers
Matilda Purdy:	Ethel Wales
Sheriff Cameron:	Lee Phelps
Hendricks:	Roy Barcroft
Ward Ackerman:	George Anderson
Logan:	Eddie Dean
Fleming:	Raphel Bennett
Stage Driver:	Jack Rockwell
Uncredited Bit Part:	Walter Long
Uncredited Bit Part:	Bob Kortman

STAGECOACH WAR

Director:	Lesley Selander
Producer:	Harry Sherman
Associate Producer:	Joseph W. Engel
Photographer:	Russell Harlan
Film Editor:	Sherman A. Rose
Music Score:	John Leipold
Music Director:	Irvin Talbot
Assistant Director:	D. M. (Derwin) Abrahams
Screenplay:	Norman Houston, Harry F. Olmstead
Distributor:	Paramount
Copyright:	July 12, 1940
Released:	July 12, 1940
Length:	61-63 minutes

Cast

Hopalong Cassidy:	William Boyd
Lucky Jenkins:	Russell Hayden
Speedy McGinnis:	Britt Wood
Shirley Chapman:	Julie Carter
Jeff Chapman:	J. Farrell MacDonald
Neal Holt:	Harvey Stephens
Smiley:	Rad Robinson
Quince Cobalt:	Eddie Waller
Twister Maxwell:	Frank Lackteen
Mart Gunther:	Jack Rockwell
Tom:	Eddie Dean
Musicians:	The King's Men
Uncredited Bit Part:	Hank Bell
Uncredited Bit Part:	Bob Kortman

THREE MEN FROM TEXAS

Director:	Lesley Selander
Producer:	Harry Sherman
Associate Producer:	Joseph W. Engel
Photographer:	Russell Harlan
Film Editor:	Carrol Lewis
Supervising Editor:	Sherman A. Rose
Music Score:	Victor Young
Assistant Director:	D. M. (Derwin) Abrahams
Screenplay:	Norton S. Parker
Distributor:	Paramount
Copyright:	November 15, 1940
Released:	November 15, 1940
Length:	70-76 minutes

Cast

Hopalong Cassidy:	William Boyd
Lucky Jenkins:	Russell Hayden
California Carlson:	Andy Clyde
Bruce Morgan:	Morris Ankrum
Capt. Andrews:	Morgan Wallace
Pico Serrano:	Thornton Edwards
Paquita:	Esther Estrella
Thompson:	Davidson Clark
Gardner:	Dick Curtis
Dave:	George Lollier
Ben Stokes:	Glenn Strange
Juanito:	Neyl Marx
Uncredited Bit Part:	Jim Corey
Uncredited Bit Part:	Bob Burns
Uncredited Bit Part:	George Morrell
Uncredited Bit Part:	Frank McCarroll
Uncredited Bit Part:	Lucio Villegas

THE DOOMED CARAVAN

Director:	Lesley Selander
Producer:	Harry Sherman
Associate Producer:	Joseph W. Engel
Photographer:	Russell Harlan
Film Editor:	Carrol Lewis
Supervising Editor:	Sherman A. Rose
Music Directors:	Irvin Talbot, John Leipold
Assistant Director:	D. M. (Derwin) Abrahams
Screenplay:	Johnston McCulley, J. Benton Cheney
Distributor:	Paramount
Copyright:	January 10, 1941
Released:	January 10, 1941
Length:	70-76 minutes

Cast

Hopalong Cassidy:	William Boyd
Lucky Jenkins:	Russell Hayden
California Jack:	Andy Clyde
Jane Travers:	Minna Gombell
Stephen Wescott:	Morris Ankrum
Diana Westcott:	Georgia Hawkins
Ed Martin:	Trevor Bardette
Jim Ferber:	Pat J. O'Brien
Pete Gregg:	Raphael Bennett
Don Pedro:	Jose Luis Tortosa
Uncredited Bit Part:	Henry Wills
Uncredited Bit Part:	Edward Cassidy
Uncredited Bit Part:	Martin Garralaga
Uncredited Bit Part:	Chick Harmon

IN OLD COLORADO

Director:	Howard Bretherton
Producer:	Harry Sherman
Associate Producer:	Joseph W. Engel
Photographer:	Russell Harlan
Film Editor:	Carrol Lewis
Supervising Editor:	Sherman A. Rose
Music Directors:	Irvin Talbot, John Leipold
Assistant Director:	D. M. (Derwin) Abrahams
Screenplay:	Norton S. Parker, J. Benton Cheney
Distributor:	Paramount
Copyright:	March 14, 1941
Released:	March 14, 1941
Length:	66-67 minutes

Cast

Hopalong Cassidy:	William Boyd
Lucky Jenkins:	Russell Hayden
California Carlson:	Andy Clyde
Myra Woods:	Margaret Hayes
Jo Weiler:	Morris Ankrum
Ma Woods:	Sarah Padden
Nosey Hawkins:	Cliff Nazarro
George Davidson:	Stanley Andrews
Hank Merritt:	James Seay
Jack Collins:	Morgan Wallace
Blackie Reed:	Weldon Heyburn
Jim Stack:	Eddy Waller
Confidence Man:	Philip Van Zandt
Uncredited Bit Part:	Curley Davidson (Dresden)
Uncredited Bit Part:	Henry Wills
Uncredited Bit Part:	Glenn Strange

BORDER VIGILANTES

Director:	Derwin Abrahams
Producer:	Harry Sherman
Associate Producer:	Joseph W. Engel
Photographer:	Russell Harlan
Film Editor:	Robert Warwick
Supervising Editor:	Sherman A. Rose
Music Directors:	Irvin Talbot, John Leipold
Assistant Director:	Frederick Spencer
Screenplay:	J. Benton Cheney
Distributor:	Paramount
Copyright:	April 18, 1941
Released:	April 18, 1941
Length:	61-63 minutes

Cast

Hopalong Cassidy:	William Boyd
Lucky Jenkins:	Russell Hayden
California Carlson:	Andy Clyde
Helen Forbes:	Frances Gifford
Henry Logan:	Victory Jory
Aunt Jennifer:	Ethel Wales
Dan Forbes:	Morris Ankrum
Jim Yager:	Tom Tyler
Ed Stone:	Hal Taliaferro
Hank Weaver:	Jack Rockwell
Lafe Willis:	Britt Wood
Wagon Driver:	Hank Worden
Banker Stevens:	Edward Earle
Liveryman:	Hank Bell
Uncredited Bit Part:	Curley Davidson (Dresden)
Uncredited Bit Part:	Al Haskell
Uncredited Bit Part:	Chuck Morrison
Uncredited Bit Part:	Ted Wells
Uncredited Bit Part:	Chick Hannon

PIRATES ON HORSEBACK

Director:	Lesley Selander
Producer:	Harry Sherman
Associate Producer:	Joseph W. Engel
Photographer:	Russell Harlan
Film Editor:	Fred Feitshans, Jr.
Supervising Editor:	Sherman A. Rose
Music Directors:	John Leipold, Maurice Lawrence
Assistant Director:	Frederick Spencer
Screenplay:	Ethel La Blanche, J. Benton Cheney
Distributor:	Paramount
Copyright:	May 23, 1941
Released:	May 31, 1941
Length:	61-63 minutes

Cast

Hopalong Cassidy:	William Boyd
Lucky Jenkins:	Russell Hayden
California Carlson:	Andy Clyde
Ace Gibson:	Morris Ankrum
Ben Pendelton:	Britt Wood
Trudy Pendleton:	Eleanor Stewart
Bill Watson:	William Haade
Jud Carter:	Dennis Moore
Sheriff John Blake:	Henry Hall
Stable Owner:	Jack Rockwell
Uncredited Bit Part:	Silver Tip Baker

WIDE OPEN TOWN

Director:	Lesley Selander
Producer:	Harry Sherman
Associate Producer:	Lewis J. Rachmil
Photographer:	Russell Harlan
Film Editor:	Carrol Lewis
Supervising Editor:	Sherman A. Rose
Music Directors:	John Leipold, Irvin Talbot
Assistant Director:	Frederick Spencer
Screenplay:	J. Benton Cheney, Harrison Jacobs
Uncredited Source:	*Hopalong Cassidy Returns* (Paramount, 1936)
Distributor:	Paramount
Copyright:	August 15, 1941
Released:	August 8, 1941
Length:	77-79 minutes

Cast

Hopalong Cassidy:	William Boyd
Lucky Jenkins:	Russell Hayden
California Carlson:	Andy Clyde
Jim Stuart:	Morris Ankrum
Joan Stuart:	Bernice Kay (Cara Williams)
Belle Langtry:	Evelyn Brent
Steve Fraser:	Victory Jory
Tom Wilson:	Kenneth Harlan

Red:	Roy Barcroft
Ed Stark:	Glenn Strange
Brad Jackson:	Edward Cassidy
Rancher:	Jack Rockwell
Blackie:	Bob Kortmann
Pete Carter:	George Cleveland

STICK TO YOUR GUNS

Director:	Lesley Selander
Producer:	Harry Sherman
Associate Producer:	Lewis J. Rachmil
Photographer:	Russell Harlan
Film Editor:	Carrol Lewis
Supervising Editor:	Sherman A. Rose
Music Director:	Irvin Talbot
Assistant Director:	Glenn Cook
Uncredited Source:	*Bar 20 Rides Again* (Paramount, 1935)
Distributor:	Paramount
Copyright:	September 19, 1941
Released:	September 27, 1941
Length:	61-63 minutes

Cast

Hopalong Cassidy:	William Boyd
California Carlson:	Andy Clyde
Johnny Nelson:	Brad King
June Winters:	Jacqueline (Jennifer) Holt
Jud Winters:	Henry Hall
Nevada:	Dick Curtis
Gila:	Weldon Heyburn
Carp:	Jack Rockwell
Elbows:	Ian McDonald
Layton:	Kermit Maynard
Long Ben:	Charles Middleton
Frenchy Smith:	Homer Holcomb
Waffles:	Tom London
Buck:	Joe Whitehead
Tex:	Jack Smith
Red:	Jack Trent
Ed:	Mickey Eissa
Uncredited Bit Part:	Robert Card
Uncredited Bit Part:	Herman Hack
Uncredited Bit Part:	Frank Ellis
Singers:	The Jimmy Wakeley Trio (Jimmy Wakely, Johnny Bond, Dick Rinehart)

SECRETS OF THE WASTELAND

Director:	Derwin Abrahams
Producer:	Harry Sherman
Associate Producer:	Lewis J. Rachmil
Photographer:	Russell Harlan
Film Editor:	Fred Feitshans, Jr.
Supervising Editor:	Sherman A. Rose
Music Score:	John Leipold
Music Director:	Irvin Talbot
Assistant Director:	John Sherwood
Screenplay:	Gerald Geraghty
Source Novel:	Bliss Lomax (Harry Sinclair Drago), *Secret of the Wastelands*, (1940)
Distributor:	Paramount
Copyright:	November 1, 1941
Released:	November 15, 1941
Length:	66 minutes

Cast

Hopalong Cassidy:	William Boyd
California Carlson:	Andy Clyde
Johnny Nelson:	Brad King
Moy Soong:	Soo Yong
Jennifer Kendall:	Barbara Britton
Slade Salters:	Douglas Fowley
Clay Elliott:	Keith Richards
Quon:	Richard Loo
Doy Kee:	Lee Tung Foo
Dr. Malcolm Birdsell:	Gordon Hart
Clanton:	Earl Gunn
Hollister:	Ian McDonald
Williams:	John Rawlins
Ying:	Roland Got
Prof. Waldo Stubbs:	Hal Price
Sheriff Mulhall:	Jack Rockwell

OUTLAWS OF THE DESERT

Director:	Howard Bretherton
Producer:	Harry Sherman
Associate Producer:	Lewis J. Rachmil
Photographer:	Russell Harlan
Film Editor:	Carrol Lewis
Supervising Editor:	Sherman A. Rose
Music Score:	John Leipold
Music Director:	Irvin Talbot
Assistant Director:	Glenn Cook
Screenplay:	J. Benton Cheney, Bernard McConville
Distributor:	Paramount
Copyright:	December 3, 1941
Released:	November 1, 1941
Length:	66 minutes

Cast

Hopalong Cassidy:	William Boyd
California Carlson:	Andy Clyde
Johnny Nelson:	Brad King
Shiek Suleiman:	Duncan Renaldo
Susan Grant:	Jean Phillips
Charles Grant:	Forrest Stanley
Mrs. Jane Grant:	Nina Guilbert
Marie Karitza:	Luli Deste
Nikki Karitza:	Albert Morin
Yussef:	George J. Lewis
Shiek Feran el Kadir:	Jean del Val
Major Crawford:	George Woolsley
Salim:	Mickey Eissa
Ali:	Jamiel Hassan

TWILIGHT ON THE TRAIL

Director:	Howard Bretherton
Producer:	Harry Sherman
Associate Producer:	Lewis J. Rachmil
Photographer:	Russell Harlan
Film Editor:	Fred Feitshans, Jr.
Supervising Editor:	Sherman A. Rose
Music Score:	John Leipold
Music Director:	Irvin Talbot
Assistant Director:	Glenn Cook
Screenplay:	J. Benton Cheney, Ellen Corby, Cecile Kramer
Distributor:	Paramount
Copyright:	December 3, 1941
Released:	September 27, 1941
Length:	58-66 minutes

Cast

Hopalong Cassidy:	William Boyd
California Carlson:	Andy Clyde
Johnny Nelson:	Brad King
Lucy Brent:	Wanda McKay
Jim Brent:	Jack Rockwell
Nat Kerby:	Norman Willis
Art Drake:	Robert Kent
Tim Gregg:	Tom London
Steve Farley:	Frank Austin
Stage Driver:	Clem Fuller
Drummer:	Johnny Powers
Uncredited Bit Part:	Bob Kortman
Uncredited Bit Part:	Bud Osborne
Uncredited Bit Part:	Frank Ellis
Singers:	The Jimmy Wakeley Trio (Jimmy Wakely, Johnny Bond, Dick Rinehart)

RIDERS OF THE TIMBERLINE

Director:	Lesley Selander
Producer:	Harry Sherman
Associate Producer:	Lewis J. Rachmil
Photographer:	Russell Harlan
Film Editor:	Fred Feitshans, Jr.
Supervising Editor:	Sherman A. Rose
Music Score:	John Leipold
Music Director:	Irvin Talbot
Assistant Director:	Glenn Cook
Screenplay:	J. Benton Cheney
Distributor:	Paramount
Copyright:	December 3, 1941
Released:	September 17, 1941
Length:	59-67 minutes

Cast

Hopalong Cassidy:	William Boyd
California Carlson:	Andy Clyde
Johnny Nelson:	Brad King
Baptiste Deschamp:	Victory Jory
Elaine Kerrigan:	Eleanor Stewart
Jim Kerrigan:	J. Farrell McDonald
Donna Ryan:	Anna Q. Nilsson
Bill Slade:	Tom Tyler
Preston Yates:	Edward Keane
Ed Petrie:	Hal Taliaferro
Larry:	Mickey Eissa
Uncredited Bit Part:	Hank Bell
Singers:	The Guardsmen

LEATHER BURNERS

Director:	Joseph A. Henabery
Producer:	Harry Sherman
Associate Producer:	Lewis J. Rachmil
Photographer:	Russell Harlan
Film Editor:	Carrol Lewis
Music Score:	Samuel Kaylin
Music Director:	Irvin Talbot
Assistant Director:	Glenn Cook
Screenplay:	Jo Pagano
Source Novel:	Bliss Lomax (Harry Sinclair Drago), *The Leather Burners* (1940)
Distributor:	United Artists
Copyright:	October 26, 1942
Released:	May 28, 1942
Length:	58-66 minutes

Cast

Hopalong Cassidy:	William Boyd
California Carlson:	Andy Clyde
Johnny Travers:	Jay Kirby
Dan Slack:	Victor Jory
Sam Bucktoe:	George Givot
Sharon Longstreet:	Shelley Spencer
Bobby Longstreet:	Bobby Larson
Harrison Brooke:	George Reeves
Lafe Bailey:	Hal Taliaferro
Bart Galey:	Forbes Murray
Uncredited Bit Parts:	Robert Mitchum
Uncredited Bit Parts:	Bob Kortman
Uncredited Bit Parts:	Herman Hack

HOPPY SERVES A WRIT

Director:	George Archainbaud
Producer:	Harry Sherman
Associate Producer:	Lewis J. Rachmil
Photographer:	Russell Harlan
Film Editor:	Sherman A. Rose
Music Director:	Irvin Talbot
Assistant Director:	Glenn Cook
Screenplay:	Gerald Geraghty
Source Novel:	Clarence E. Mulford, *Hopalong Cassidy Serves A Writ* (1941)
Distributor:	United Artists
Copyright:	November 25, 1942
Released:	March 12, 1942
Length:	66-67 minutes

Cast

Hopalong Cassidy:	William Boyd
California Carlson:	Andy Clyde
Johnny Travers:	Jay Kirby
Tom Jordan:	Victor Jory
Steve Jordan:	George Reeves
Greg Jordan:	Hal Taliaferro
Jean Hollister:	Jan Christy
Ben Hollister:	Forbes Murray
Rigney:	Robert Mitchum
Danvers:	Bryon Foulger
Jim Belson:	Earle Hodgins
Tom Colby:	Roy Barcroft
Card Player:	Ben Corbett
Uncredited Bit Parts:	Art Mix

UNDERCOVER MAN

Director:	Lesley Selander
Producer:	Harry Sherman
Associate Producer:	Lewis J. Rachmil
Photographer:	Russell Harlan
Film Editor:	Carrol Lewis
Supervising Editor:	Sherman A. Rose
Music Director:	Irvin Talbot
Assistant Director:	Glenn Cook
Screenplay:	J. Benton Cheney
Distributor:	United Artists
Copyright:	December 10, 1942
Released:	October 23, 1942
Length:	58-68 minutes

Cast

Hopalong Cassidy:	William Boyd
California Carlson:	Andy Clyde
Breezy Travers:	Jay Kirby
Don Tomas Gonzales:	Antonio Moreno
Louise Saunders:	Nora Lane
Miguel:	Chris-Pin Martin
Dolores Gonzales:	Esther Estrella
Ed Carson/Idaho Pete Jackson:	John Vosper
Rosita Lopez:	Eva Puig
Bob Saunders:	Alan Baldwin
Capt. John Hawkins:	Jack Rockwell
Bert:	Pierce Lyden
Chavez:	Tony Roux
Jim:	Ted Wells
Cortez:	Martin Garralaga
Caballero:	Joe Dominguez
Sheriff Blackburn:	Earle Hodgins
Uncredited Bit Parts:	Bennett George
Uncredited Bit Parts:	Frank Ellis

BORDER PATROL

Director:	Lesley Selander
Producer:	Harry Sherman
Associate Producer:	Lewis J. Rachmil
Photographer:	Russell Harlan
Film Editor:	Sherman A. Rose
Music Director:	Irvin Talbot
Assistant Director:	Glenn Cook
Screenplay:	Michael Wilson
Distributor:	United Artists
Copyright:	December 11, 1942
Released:	April 2, 1943
Length:	60-65 minutes

Cast

Hopalong Cassidy:	William Boyd
California Carlson:	Andy Clyde
Johnny Travers:	Jay Kirby
Orestes Krebs:	Russell Simpson
Inez La Baca:	Claudia Drake
Don Enrique Perez:	George Reeves
Commandant:	Duncan Renaldo
Loren:	Pierce Lyden
Quinn:	Robert Mitchum
Barton:	Cliff Parkinson
Cook:	Earle Hodgins
Uncredited Bit Parts:	Merrill McCormack

LOST CANYON

Director:	Lesley Selander
Producer:	Harry Sherman
Associate Producer:	Lewis J. Rachmil
Photographer:	Russell Harlan
Film Editor:	Carrol Lewis
Music Director:	Irvin Talbot
Assistant Director:	Glenn Cook
Screenplay:	Harry O. Hoyt
Uncredited Source:	*Rustler's Valley* (Paramount, 1937)
Distributor:	United Artists
Copyright:	December 11, 1942
Released:	December 18, 1943
Length:	61 minutes

Cast

Hopalong Cassidy:	William Boyd
California Carlson:	Andy Clyde
Johnny Travers:	Jay Kirby
Laura Clark:	Lola Lane
Jeff Burton:	Douglas Fowley
Tom Clark:	Herbert Rawlinson

Zack Rogers:	Guy Usher
Wade Haskell:	Karl Hackett
Jim Stanton:	Hugh Prosser
Joe:	Bob Kortman
Uncredited Bit Parts:	Merrill McCormack
Uncredited Bit Parts:	Keith Richars
Uncredited Bit Parts:	Herman Hack
Uncredited Bit Parts:	George Morrell
Uncredited Bit Parts:	Spade Cooley
Uncredited Bit Parts:	John Cason
Uncredited Bit Parts:	Henry Wills
Singers:	The Sportsmen Quartette

COLT COMRADES

Director:	Lesley Selander
Producer:	Harry Sherman
Associate Producer:	Lewis J. Rachmil
Photographer:	Russell Harlan
Film Editor:	Fred W. Berger
Music Director:	Irvin Talbot
Assistant Director:	Glenn Cook
Screenplay:	Michael Wilson
Source Novel:	Bliss Lomax (Harry Sinclair Drago), *Colt Comrades* (1939)
Distributor:	United Artists
Copyright:	February 1, 1943
Released:	June 18, 1943
Length:	67 minutes

Cast

Hopalong Cassidy:	William Boyd
California Carlson:	Andy Clyde
Johnny Travers:	Jay Kirby
Lucy Whitlock:	Lois Sherman
Lin Whitlock:	George Reeves
Jeb Hardin:	Victory Jory
Joe Brass:	Douglas Fowley
Varney:	Herbert Rawlinson
Wildcat Willy:	Earle Hodgins
Dirk Mason:	Robert Mitchum
Sheriff:	Russell Simpson
Postmaster:	Jack Mulhall
Uncredited Bit Parts:	Fred Kohler, Jr.

BAR 20

Director:	Lesley Selander
Producer:	Harry Sherman
Associate Producer:	Lewis J. Rachmil
Photographer:	Russell Harlan
Film Editor:	Carrol Lewis
Music Director:	Irvin Talbot
Assistant Director:	Glenn Cook
Screenplay:	Morton Grant, Norman Houston, Michael Wilson
Distributor:	United Artists
Copyright:	June 1, 1943
Released:	October 1, 1943
Length:	54 minutes

Cast

Hopalong Cassidy:	William Boyd
California Carlson:	Andy Clyde
Lin Bradley:	George Reeves
Mark Jackson:	Victory Jory
Slash:	Douglas Fowley
Tom:	Earle Hodgins
Richard Adams:	Robert Mitchum
Marie Stevens:	Dustine Farnum
Mrs. Stevens:	Betty Blythe
Quirt Rankin:	Francis McDonald
Uncredited Bit Parts:	Buck Bucko

FALSE COLORS

Director:	George Archainbaud
Producer:	Harry Sherman
Associate Producer:	Lewis J. Rachmil
Photographer:	Russell Harlan
Film Editor:	Fred W. Berger
Supervising Editor:	Carrol Lewis
Music Director:	Irvin Talbot
Assistant Director:	Glenn Cook
Screenplay:	Bennett Cohen

Distributor:	United Artists
Copyright:	September 27, 1943
Released:	November 5, 1943
Length:	64-65 minutes

Cast

Hopalong Cassidy:	William Boyd
California Carlson:	Andy Clyde
Jimmy Rogers:	Jimmy Rogers
Mark Foster:	Douglas Dumbrille
Bud Lawton/Kit Moyer:	Tom Seidel
Faith Lawton:	Claudia Drake
Sonora:	Glenn Strange
Rip Austin:	Robert Mitchum
Lefty:	Pierce Lyden
Sheriff Clem Martin:	Roy Barcroft
Jay Griffin:	Earle Hodgins
Judge Stevens:	Sam Flint
Uncredited Bit Parts:	Tom London
Uncredited Bit Parts:	Dan White
Uncredited Bit Parts:	Elmer Jerome
Uncredited Bit Parts:	George Morrell
Uncredited Bit Parts:	Ray Jones
Uncredited Bit Parts:	Bob Burns
Uncredited Bit Parts:	Tom Smith

RIDERS OF THE DEADLINE

Director:	Lesley Selander
Producer:	Harry Sherman
Associate Producer:	Lewis J. Rachmil
Photographer:	Russell Harlan
Film Editor:	Walter Hannemann
Supervising Editor:	Carrol Lewis
Music Director:	Irvin Talbot
Assistant Director:	Glenn Cook
Screenplay:	Bennett Cohen
Uncredited Source:	*The Desert Bandit* (Republic, 1941)
Distributor:	United Artists
Copyright:	September 27, 1943
Released:	December 3, 1943
Length:	60-70 minutes

Cast

Hopalong Cassidy:	William Boyd
California Carlson:	Andy Clyde
Jimmy Rogers:	Jimmy Rogers
Sue Mason:	Frances Woodward
Simon Crandall:	William Halligan
Nick Drago:	Robert Mitchum
Tim Mason:	Richard Crane
Gwiner Madigan:	Anthony Warde
Deputy Sheriff Martin:	Hugh Prosser
Captain Jennings:	Herbert Rawlinson
Tex:	Jack Rockwell
Sourdough:	Earle Hodgins
Private Calhoun:	Montie Montana
Uncredited Bit Parts:	Jim Bannon
Uncredited Bit Parts:	Bill Beckford
Uncredited Bit Parts:	Pierce Lyden
Uncredited Bit Parts:	Art Felix
Uncredited Bit Parts:	Roy Bucko
Uncredited Bit Parts:	Cliff Parkinson

TEX MASQUERADE

Director:	George Archainbaud
Producer:	Harry Sherman
Associate Producer:	Lewis J. Rachmil
Photographer:	Russell Harlan
Film Editor:	Walter Hannemann
Supervising Editor:	Carrol Lewis
Music Director:	Irvin Talbot
Assistant Director:	Glenn Cook
Screenplay:	Jack Lait, Jr., Norman Houston
Distributor:	United Artists
Copyright:	December 8, 1943
Released:	February 8, 1943
Length:	58-59 minutes

Cast

Hopalong Cassidy:	William Boyd
California Carlson:	Andy Clyde
Jimmy Rogers:	Jimmy Rogers
Ace Maxson:	Don Castello

Virginia Curtis:	Mady Correll
Sam Nolan:	Francis McDonald
J. K. Trimble:	Russell Simpson
John Martindale:	J. Farrell MacDonald
Mrs. Martindale:	June Terry Pickrell
James Corwin:	Nelson Leigh
Constable Rowbottom:	Robert McKenzie
Al:	Pierce Lyden
Lew Sykes:	Bill Hunter
Jeff:	John Merton
Uncredited Bit Parts:	Keith Richards
Uncredited Bit Parts:	George Morrell

LUMBERBACK

Director:	Lesley Selander
Producer:	Harry Sherman
Associate Producer:	Lewis J. Rachmil
Photographer:	Russell Harlan
Film Editor:	Fred W. Berger
Supervising Editor:	Carrol Lewis
Music Director:	Irvin Talbot
Assistant Director:	Glenn Cook
Screenplay:	Norman Houston, Barry Shipman
Distributor:	United Artists
Copyright:	March 24, 1944
Released:	April 28, 1944
Length:	65 minutes

Cast

Hopalong Cassidy:	William Boyd
California Carlson:	Andy Clyde
Jimmy Rogers:	Jimmy Rogers
Daniel J. Keefer:	Douglass Dumbrille
Julie Peters Jordan:	Ellen Hall
Benton C. Jordan:	John Whitney
Clyde Fenwick:	Francis McDonald
Aunt Abbie Peters:	Ethel Wales
Buck Peters:	Herbert Rawlinson
Taggart:	Hal Taliaferro
Big Joe Williams:	Charles Morton
Mrs. Williams:	Frances Morris
Sheriff Miles:	Jack Rockwell
Uncredited Bit Parts:	Henry Wills
Uncredited Bit Parts:	Bob Burns
Uncredited Bit Parts:	Pierce Lyden
Uncredited Bit Parts:	Hank Worden
Uncredited Bit Parts:	Earle Hodgins

MYSTERY MAN

Director:	George Archainbaud
Producer:	Harry Sherman
Associate Producer:	Lewis J. Rachmil
Photographer:	Russell Harlan
Film Editor:	Fred W. Berger
Supervising Editor:	Carrol Lewis
Music Director:	Irvin Talbot
Assistant Director:	Glenn Cook
Screenplay:	J. Benton Cheney
Distributor:	United Artists
Copyright:	May 31, 1944
Released:	May 31, 1944
Length:	58 minutes

Cast

Hopalong Cassidy:	William Boyd
California Carlson:	Andy Clyde
Jimmy Rogers:	Jimmy Rogers
Bud Trilling:	Don Castello
Diane Newhall:	Eleanor Stewart
Sheriff Sam Newhall:	Forrest Taylor
Bert Ragan:	Francis McDonald
Tad Blane:	Jack Rockwell
Bill:	John Merton
Red:	Pierce Lyden
Tom Hanlon:	Bob Burns
Singing Ranch Hand:	Ozie Waters
Uncredited Bit Parts:	Bill Hunter
Uncredited Bit Parts:	Art Mix
Uncredited Bit Parts:	Bob Barker
Uncredited Bit Parts:	Hank Bell
Uncredited Bit Parts:	George Morell

FORTY THIEVES

Director:	Lesley Selander
Producer:	Harry Sherman
Associate Producer:	Lewis J. Rachmil
Photographer:	Russell Harlan
Film Editor:	Carrol Lewis
Music Score:	Mort Glickman
Music Supervisor:	David Chudnow
Assistant Director:	George Tobin
Screenplay:	Michael Wilson, Bernie Kamins
Distributor:	United Artists
Copyright:	June 23, 1944
Released:	June 23, 1944
Length:	60 minutes

Cast

Hopalong Cassidy:	William Boyd
California Carlson:	Andy Clyde
Jimmy Rogers:	Jimmy Rogers
Tad Hammond:	Douglass Dumbrille
Katherine Reynolds:	Louise Currie
Judge Reynolds:	Robert Frazier
Jerry Doyle:	Kirk Alyn
Buck Peters:	Herbert Rawlinson
Ike Simmons:	Glenn Strange
Clanton:	Hal Taliaferro
Sam Garms:	Jack Rockwell
Joe Garms:	Bob Kortman
Uncredited Bit Parts:	Earle Hodgins

THE DEVIL'S PLAYGROUND

Director:	George Archainbaud
Producer:	Lewis J. Rachmil
Executive Producer:	William Boyd
Photographer:	Mack Stengler
Film Editor:	Fred W. Berger
Music:	David Chudnow
Assistant Director:	George Tobin
Screenplay:	Ted Wilson
Distributor:	United Artists
Copyright:	November 15, 1946
Released:	November 15, 1946
Length:	62-65 minutes

Cast

Hopalong Cassidy:	William Boyd
California Carlson:	Andy Clyde
Lucky Jenkins:	Rand Brooks
Mrs. Evans:	Elaine Riley
Judge Morton:	Robert Elliott
Sheriff:	Joseph J. Greene
Roberts:	Francis McDonald
Curly Evans:	Ned Young
Dan'l:	Earle Hodgins
U. S. Marshal:	George Eldredge
Wolfe:	Everett Shields
Shorty:	John George
Uncredited Bit Parts:	Glenn Strange

FOOLS' GOLD

Director:	George Archainbaud
Producer:	Lewis J. Rachmil
Executive Producer:	William Boyd
Photographer:	Mack Stengler
Film Editor:	Fred W. Berger
Music:	David Chudnow
Assistant Director:	George Tobin
Screenplay:	Doris Schroeder
Distributor:	United Artists
Copyright:	January 31, 1947
Released:	January 31, 1947
Length:	63-65 minutes

Cast

Hopalong Cassidy:	William Boyd
California Carlson:	Andy Clyde
Lucky Jenkins:	Rand Brooks
Professor Dixon:	Robert Emmett Keane
Jessie Dixon:	Jane Randolph
Bruce Landy:	Stephen Barclay

Col. Jed Landy:	Forbes Murray
Duke:	Harry Cording
Sandler:	Earle Hodgins
Barton:	Bob Bentley
Blackie:	William "Wee Willie" Davis
Lieutenant Anderson:	Glen B. Gallagher
Sergeant:	Ben Corbett
Speed:	Fred "Snowflake" Toones

UNEXPECTED GUEST

Director:	George Archainbaud
Producer:	Lewis J. Rachmil
Executive Producer:	William Boyd
Photographer:	Mack Stengler
Film Editor:	Fred W. Berger
Music:	David Chudnow
Assistant Director:	George Tobin
Screenplay:	Ande Lamb
Distributor:	United Artists
Copyright:	March 28, 1947
Released:	March 28, 1947
Length:	59-61 minutes

Cast

Hopalong Cassidy:	William Boyd
California Carlson:	Andy Clyde
Lucky Jenkins:	Rand Brooks
Matilda Hackett:	Una O'Connor
David J. Potter:	John Parrish
Ruth Baxter:	Patricia Tate
Ralph Baxter:	Ned Young
Joshua Colter:	Earle Hodgins
Phineas Phipps:	Joel Friedkin
Matt Ogden:	Robert B. Williams
Sheriff:	William Ruhl

DANGEROUS VENTURE

Director:	George Archainbaud
Producer:	Lewis J. Rachmil
Executive Producer:	William Boyd
Photographer:	Mack Stengler
Film Editor:	Fred W. Berger
Music:	David Chudnow
Assistant Director:	George Tobin
Screenplay:	Doris Schroeder
Distributor:	United Artists
Copyright:	May 23, 1947
Released:	May 23, 1947
Length:	59 minutes

Cast

Hopalong Cassidy:	William Boyd
California Carlson:	Andy Clyde
Lucky Jenkins:	Rand Brooks
Xeoli:	Fritz Leiber
Dr. Atwood:	Douglas Evans
Dan Morgan:	Harry Cording
Dr. Sue Harmon:	Betty Alexander
Kane:	Francis McDonald
Jose:	Neyle Morrow
Tolu:	Patricia Tate
Stark:	Bob Faust
Red:	Ken Tobey
Marshal:	Jack Quinn
Pete:	Bill Nestell

HOPPY'S HOLIDAY

Director:	George Archainbaud
Producer:	Lewis J. Rachmil
Executive Producer:	William Boyd
Photographer:	Mack Stengler
Film Editor:	Fred W. Berger
Music:	David Chudnow
Assistant Director:	George Tobin
Screenplay:	J. Benton Cheney, Bennett Cohen, Ande Lamb
Story:	Ellen Corby, Cecile Kramer
Distributor:	United Artists
Copyright:	July 19, 1947
Released:	July 18, 1947
Length:	60-70 minutes

Cast

Hopalong Cassidy:	William Boyd
California Carlson:	Andy Clyde
Lucky Jenkins:	Rand Brooks
Mayor Frank Patton:	Andrew Tombes
Gloria Patton:	Mary Ware
Dunning:	Leonard Penn
Jed:	Jeff Corey
Sheriff:	Donald Kirke
Ace:	Hollis Bane (Mike Ragan)
Jay:	Gil Patric
Bart:	Frank Henry

THE MARAUDERS

Director:	George Archainbaud
Producer:	Lewis J. Rachmil
Executive Producer:	William Boyd
Photographer:	Mack Stengler
Film Editor:	Fred W. Berger
Music:	Ralph Stanley
Assistant Director:	George Templeton
Screenplay:	Charles Belden
Distributor:	United Artists
Copyright:	September 12, 1947
Released:	July 1, 1947
Length:	63 minutes

Cast

Hopalong Cassidy:	William Boyd
California Carlson:	Andy Clyde
Lucky Jenkins:	Rand Brooks
Deacon Edwin Black:	Ian Wolfe
Susan Crowell:	Dorinda Clifton
Mrs. Crowell:	Mary Newton
Riker:	Harry Cording
County Clerk:	Earle Hodgins
Oil Driller:	Dick Bailey
Uncredited Bit Part:	Richard Alexander
Uncredited Bit Part:	Herman Hack

SILENT CONFLICT

Director:	George Archainbaud
Producer:	Lewis J. Rachmil
Executive Producer:	William Boyd
Photographer:	Mack Stengler
Film Editor:	Fred W. Berger
Music:	Darrell Calker
Assistant Director:	William Dario Faralla
Screenplay:	Charles Belden
Distributor:	United Artists
Copyright:	March 19, 1948
Released:	March 19, 1948
Length:	61 minutes

Cast

Hopalong Cassidy:	William Boyd
California Carlson:	Andy Clyde
Lucky Jenkins:	Rand Brooks
Renee Richards:	Virginia Belmont
Doc Richards:	Earle Hodgins
Speed Blaney:	James Harrison
Randall:	Forbes Murray
Clerk:	John Butler
Jake:	Herbert Rawlinson
Rancher:	Richard Alexander
Rancher:	Don Haggerty

THE DEAD DON'T DREAM

Director:	George Archainbaud
Producer:	Lewis J. Rachmil
Executive Producer:	William Boyd
Photographer:	Mack Stengler
Film Editor:	Fred W. Berger
Music:	Darrell Calker
Assistant Director:	William Dario Faralla
Screenplay:	Francis Rosenwald
Distributor:	United Artists
Copyright:	April 30, 1948
Released:	April 30, 1948
Length:	62-68 minutes

Cast

Hopalong Cassidy:	William Boyd
California Carlson:	Andy Clyde
Lucky Jenkins:	Rand Brooks
Jeff Potter:	John Parrish
Larry Potter:	Bob Gabriel
Earl Wesson:	Leonard Penn
Mary Benton:	Mary Tucker
Bart Lansing:	Francis McDonald
Duke:	Richard Alexander
Jesse Williams:	Stanley Andrews
Sheriff Thompson:	Forbes Murray
Deputy:	Don Haggerty

SINISTER JOURNEY

Director:	George Archainbaud
Producer:	Lewis J. Rachmil
Executive Producer:	William Boyd
Photographer:	Mack Stengler
Film Editor:	Fred W. Berger
Music:	Darrell Calker
Assistant Director:	William Dario Faralla
Screenplay:	Doris Schroeder
Distributor:	United Artists
Copyright:	June 11, 1948
Released:	June 11, 1948
Length:	59-72 minutes

Cast

Hopalong Cassidy:	William Boyd
California Carlson:	Andy Clyde
Lucky Jenkins:	Rand Brooks
Jessie Garvin:	Elaine Riley
Lee Garvin:	John Kellogg
Harmon Roberts:	Don Haggerty
Tom Smith:	Stanley Andrews
Banks:	Harry Strang
Storekeeper:	Herbert Rawlinson
Constable Reardon:	John Butler
Ben Watts:	Will Orlean
Engineer:	Wayne C. Tredway

BORROWED TROUBLE

Director:	George Archainbaud
Producer:	Lewis J. Rachmil
Executive Producer:	William Boyd
Photographer:	Mack Stengler
Film Editor:	Fred W. Berger
Music:	Darrell Calker
Assistant Director:	William Dario Faralla
Screenplay:	Charles Belden
Distributor:	United Artists
Copyright:	July 1, 1948
Released:	July 1, 1948
Length:	58-61 minutes

Cast

Hopalong Cassidy:	William Boyd
California Carlson:	Andy Clyde
Lucky Jenkins:	Rand Brooks
Lucy Abbott:	Anne O'Neal
Steve Mawson:	John Parrish
Dink Davis:	Cliff Clark
Lola Blair:	Helen Chapman
Sheriff:	Earle Hodgins
Groves:	Herbert Rawlinson
Lippy:	Don Haggerty
Rocky:	James Harrison
Henchman:	Clark Stevens
Henchman:	George Sowards
Child:	Eileen Janssen
Child:	Nancy Stone
Child:	Jimmy Crane
Child:	Billy O'Leary
Child:	Norman Ollestad, Jr.

FALSE PARADISE

Director:	George Archainbaud
Producer:	Lewis J. Rachmil
Executive Producer:	William Boyd
Photographer:	Mack Stengler
Film Editor:	Fred W. Berger
Music:	Ralph Stanley
Assistant Director:	William Dario Faralla
Screenplay:	Harrison Jacobs
Distributor:	United Artists
Copyright:	July 15, 1948
Released:	September 10, 1948
Length:	59-66 minutes

Cast

Hopalong Cassidy:	William Boyd
California Carlson:	Andy Clyde
Lucky Jenkins:	Rand Brooks
Anne Larson:	Elaine Riley
Waite:	Cliff Clark
Prof. Alonzo Larson:	Joel Friedkin
Bentley:	Kenneth MacDonald
Deal Marden:	Don Haggerty
Radley:	George Eldredge
Sam:	Richard Alexander
Buck:	Zon Murray
Uncredited Bit Part:	William Norton Bailey

STRANGE GAMBLE

Director:	George Archainbaud
Producer:	Lewis J. Rachmil
Executive Producer:	William Boyd
Photographer:	Mack Stengler
Film Editor:	Fred W. Berger
Music:	Ralph Stanley
Assistant Director:	William Dario Faralla
Screenplay:	Doris Schroeder
Distributor:	United Artists
Copyright:	August 1, 1948
Released:	October 8, 1948
Length:	61-62 minutes

Cast

Hopalong Cassidy:	William Boyd
California Carlson:	Andy Clyde
Lucky Jenkins:	Rand Brooks
Nora Murray:	Elaine Riley
Mary Murray:	Joan Barton
Ora Mordigan:	James Craven
Pete:	Robert B. Williams
DeLara:	Albert Morin
Doc White:	Joel Friedkin
John Murray:	Herbert Rawlinson
Longhorn:	Francis McDonald
Sid Murray:	William Leicester
Wong:	Lee Tung Foo

Appendix 3
TELEVISION FILMOGRAPHY, TRAILERS, AND STATION LISTINGS

TV SERIES-OGRAPHY

THE DEVIL'S PLAY-GROUND
#55

Director:	George Archainbaud
Producer:	Lewis J. Rachmil
Executive Producer:	William Boyd
Photographer:	Mack Stengler
Film Editor:	Fred W. Berger
Music:	David Chudnow
Assistant Director:	George Tobin
Screenplay:	Ted Wilson
Copyright:	November 15, 1946

Cast

Hopalong Cassidy:	William Boyd
California Carlson:	Andy Clyde
Lucky Jenkins:	Rand Brooks
Mrs. Evans:	Elaine Riley
Judge Morton:	Robert Elliott
Sheriff:	Joseph J. Greene
Roberts:	Francis McDonald
Curly Evans:	Ned Young
Dan'l:	Earle Hodgins
U. S. Marshal:	George Eldredge
Wolfe:	Everett Shields
Shorty:	John George
Uncredited Bit Parts:	Glenn Strange

FOOLS' GOLD
#56

Director:	George Archainbaud
Producer:	Lewis J. Rachmil
Executive Producer:	William Boyd
Photographer:	Mack Stengler
Film Editor:	Fred W. Berger
Music:	David Chudnow
Assistant Director:	George Tobin
Screenplay:	Doris Schroeder
Copyright:	January 31, 1947

Cast

Hopalong Cassidy:	William Boyd
California Carlson:	Andy Clyde
Lucky Jenkins:	Rand Brooks
Professor Dixon:	Robert Emmett Keane
Jessie Dixon:	Jane Randolph
Bruce Landy:	Stephen Barclay
Col. Jed Landy:	Forbes Murray
Duke:	Harry Cording
Sandler:	Earle Hodgins
Barton:	Bob Bentley
Blackie:	William "Wee Willie" Davis
Lieutenant Anderson:	Glen B. Gallagher
Sergeant:	Ben Corbett
Speed:	Fred "Snowflake" Toones

DANGEROUS VENTURE
#57

Director:	George Archainbaud
Producer:	Lewis J. Rachmil
Executive Producer:	William Boyd
Photographer:	Mack Stengler
Film Editor:	Fred W. Berger
Music:	David Chudnow
Assistant Director:	George Tobin
Screenplay:	Doris Schroeder
Copyright:	May 23, 1947

Cast

Hopalong Cassidy:	William Boyd
California Carlson:	Andy Clyde
Lucky Jenkins:	Rand Brooks
Xeoli:	Fritz Leiber
Dr. Atwood:	Douglas Evans
Dan Morgan:	Harry Cording
Dr. Sue Harmon:	Betty Alexander
Kane:	Francis McDonald
Jose:	Neyle Morrow
Tolu:	Patricia Tate
Stark:	Bob Faust
Red:	Ken Tobey
Marshal:	Jack Quinn
Pete:	Bill Nestell

UNEXPECTED GUEST
#58

Director:	George Archainbaud
Producer:	Lewis J. Rachmil
Executive Producer:	William Boyd
Photographer:	Mack Stengler
Film Editor:	Fred W. Berger
Music:	David Chudnow
Assistant Director:	George Tobin
Screenplay:	Ande Lamb
Copyright:	March 28, 1947

Cast

Hopalong Cassidy:	William Boyd
California Carlson:	Andy Clyde
Lucky Jenkins:	Rand Brooks
Matilda Hackett:	Una O' Connor
David J. Potter:	John Parrish
Ruth Baxter:	Patricia Tate
Ralph Baxter:	Ned Young
Joshua Colter:	Earle Hodgins
Phineas Phipps:	Joel Friedkin
Matt Ogden:	Robert B. Williams
Sheriff:	William Ruhl

HOPPY'S HOLIDAY
#59

Director:	George Archainbaud
Producer:	Lewis J. Rachmil
Executive Producer:	William Boyd
Photographer:	Mack Stengler

Film Editor:	Fred W. Berger
Music:	David Chudnow
Assistant Director:	George Tobin
Screenplay:	J. Benton Cheney, Bennett Cohen, Ande Lamb
Story:	Ellen Corby, Cecile Kramer
Copyright:	July 19, 1947

Cast

Hopalong Cassidy:	William Boyd
California Carlson:	Andy Clyde
Lucky Jenkins:	Rand Brooks
Mayor Frank Patton:	Andrew Tombes
Gloria Patton:	Mary Ware
Dunning:	Leonard Penn
Jed:	Jeff Corey
Sheriff:	Donald Kirke
Ace:	Hollis Bane (Mike Ragan)
Jay:	Gil Patric
Bart:	Frank Henry

THE MARAUDERS
#60

Director:	George Archainbaud
Producer:	Lewis J. Rachmil
Executive Producer:	William Boyd
Photographer:	Mack Stengler
Film Editor:	Fred W. Berger
Music:	Ralph Stanley
Assistant Director:	George Templeton
Screenplay:	Charles Belden
Copyright:	September 12, 1947

Cast

Hopalong Cassidy:	William Boyd
California Carlson:	Andy Clyde
Lucky Jenkins:	Rand Brooks
Deacon Edwin Black:	Ian Wolfe
Susan Crowell:	Dorinda Clifton
Mrs. Crowell:	Mary Newton
Riker:	Harry Cording
County Clerk:	Earle Hodgins
Oil Driller:	Dick Bailey
Uncredited Bit Part:	Richard Alexander
Uncredited Bit Part:	Herman Hack

SILENT CONFLICT
#61

Director:	George Archainbaud
Producer:	Lewis J. Rachmil
Executive Producer:	William Boyd
Photographer:	Mack Stengler
Film Editor:	Fred W. Berger
Music:	Darrell Calker
Assistant Director:	William Dario Faralla
Screenplay:	Charles Belden
Copyright:	March 19, 1948

Cast

Hopalong Cassidy:	William Boyd
California Carlson:	Andy Clyde
Lucky Jenkins:	Rand Brooks
Renee Richards:	Virginia Belmont
Doc Richards:	Earle Hodgins
Speed Blaney:	James Harrison
Randall:	Forbes Murray
Clerk:	John Butler
Jake:	Herbert Rawlinson
Rancher:	Richard Alexander
Rancher:	Don Haggerty

THE DEAD DON'T DREAM
#62

Director:	George Archainbaud
Producer:	Lewis J. Rachmil
Executive Producer:	William Boyd
Photographer:	Mack Stengler
Film Editor:	Fred W. Berger
Music:	Darrell Calker
Assistant Director:	William Dario Faralla
Screenplay:	Francis Rosenwald
Copyright:	April 30, 1948

Cast

Hopalong Cassidy:	William Boyd
California Carlson:	Andy Clyde
Lucky Jenkins:	Rand Brooks
Jeff Potter:	John Parrish
Larry Potter:	Bob Gabriel
Earl Wesson:	Leonard Penn
Mary Benton:	Mary Tucker
Bart Lansing:	Francis McDonald
Duke:	Richard Alexander
Jesse Williams:	Stanley Andrews
Sheriff Thompson:	Forbes Murray
Deputy:	Don Haggerty

SINISTER JOURNEY
#63

Director:	George Archainbaud
Producer:	Lewis J. Rachmil
Executive Producer:	William Boyd
Photographer:	Mack Stengler
Film Editor:	Fred W. Berger
Music:	Darrell Calker
Assistant Director:	William Dario Faralla
Screenplay:	Doris Schroeder
Copyright:	June 11, 1948

Cast

Hopalong Cassidy:	William Boyd
California Carlson:	Andy Clyde
Lucky Jenkins:	Rand Brooks
Jessie Garvin:	Elaine Riley
Lee Garvin:	John Kellogg

Harmon Roberts:	Don Haggerty
Tom Smith:	Stanley Andrews
Banks:	Harry Strang
Storekeeper:	Herbert Rawlinson
Constable Reardon:	John Butler
Ben Watts:	Will Orlean
Engineer:	Wayne C. Tredway

BORROWED TROUBLE
#64

Director:	George Archainbaud
Producer:	Lewis J. Rachmil
Executive Producer:	William Boyd
Photographer:	Mack Stengler
Film Editor:	Fred W. Berger
Music:	Darrell Calker
Assistant Director:	William Dario Faralla
Screenplay:	Charles Belden
Copyright:	July 1, 1948

Cast

Hopalong Cassidy:	William Boyd
California Carlson:	Andy Clyde
Lucky Jenkins:	Rand Brooks
Lucy Abbott:	Anne O'Neal
Steve Mawson:	John Parrish
Dink Davis:	Cliff Clark
Lola Blair:	Helen Chapman
Sheriff:	Earle Hodgins
Groves:	Herbert Rawlinson
Lippy:	Don Haggerty
Rocky:	James Harrison
Henchman:	Clark Stevens
Henchman:	George Sowards
Child:	Eileen Janssen
Child:	Nancy Stone
Child:	Jimmy Crane
Child:	Billy O'Leary
Child:	Norman Ollestad, Jr.

FALSE PARADISE
#65

Director:	George Archainbaud
Producer:	Lewis J. Rachmil
Executive Producer:	William Boyd
Photographer:	Mack Stengler
Film Editor:	Fred W. Berger
Music:	Ralph Stanley
Assistant Director:	William Dario Faralla
Screenplay:	Harrison Jacobs
Copyright:	July 15, 1948

Cast

Hopalong Cassidy:	William Boyd
California Carlson:	Andy Clyde
Lucky Jenkins:	Rand Brooks
Anne Larson:	Elaine Riley
Waite:	Cliff Clark
Prof. Alonzo Larson:	Joel Friedkin
Bentley:	Kenneth MacDonald
Deal Marden:	Don Haggerty
Radley:	George Eldredge
Sam:	Richard Alexander
Buck:	Zon Murray
Uncredited Bit Part: Bailey	William Norton

STRANGE GAMBLE
#66

Director:	George Archainbaud
Producer:	Lewis J. Rachmil
Executive Producer:	William Boyd
Photographer:	Mack Stengler
Film Editor:	Fred W. Berger
Music:	Ralph Stanley
Assistant Director:	William Dario Faralla
Screenplay:	Doris Schroeder
Copyright:	August 1, 1948

Cast

Hopalong Cassidy:	William Boyd
California Carlson:	Andy Clyde
Lucky Jenkins:	Rand Brooks
Nora Murray:	Elaine Riley
Mary Murray:	Joan Barton
Ora Mordigan:	James Craven
Pete:	Robert B. Williams
DeLara:	Albert Mrin
Doc White:	Joel Friedkin
John Murray:	Herbert Rawlinson
Longhorn:	Francis McDonald
Sid Murray:	William Leicester
Wong:	Lee Tung Foo

THE KNIFE OF CARLOS VALERO
#67

Director:	Derwin Abbe
Teleplay:	Harrison Jacobs
Production Supervisor:	Glenn Cook
Film Editor:	Fred Berger
Music Director:	Raoul Kraushaar
Copyright:	September 25, 1952
Season:	First, 1952-1953

Cast

Hopalong Cassidy:	William Boyd
Red Conners:	Edgar Buchanan
Bailey/Blake:	Harry Cording
Ross:	John Crawford
Sheriff:	Olin Howlin
Carlos Valero:	Victor Millan
Trini:	Lillian Moliere
Hotel Clerk:	Bryon Foulger

ALIEN RANGE
#68

Director:	Thomas Carr
Teleplay:	Sherman L. Lowe
Production Supervisor:	Glenn Cook
Cameraman:	Mack Stengler
Film Editor:	Fred Berger
Music Director:	Raoul Kraushaar
Composers:	Nacio Herb Brown and L. Wolfe Gilbert, *Hopalong Cassidy March*
Copyright:	October 1, 1952
Season:	First, 1952-1953

Cast

Hopalong Cassidy:	William Boyd
Red Conners:	Edgar Buchanan
Jon Vandermeer:	Otto Waldis
Lilli Vandermeer:	Maria Palmer
Simon Cosgrove:	Glenn Strange
Roscoe Hicks:	James Griffith
Robbins:	Larry Hudson

BLACK SHEEP
#69

Director:	Derwin Abbe
Teleplay:	Harrison Jacobs
Production Supervisor:	Glenn Cook
Cameraman:	Mack Stengler
Music Director:	Raoul Kraushaar
Copyright:	December 26, 1952
Season:	First, 1952-1953

Cast

Hopalong Cassidy:	William Boyd
Red Conners:	Edgar Buchanan
Bob Norman:	Richard Crane
Rance Barlow:	Richard Travis
Lucy Barlow:	Claire Carleton
Mary Warner:	Antoinette Gerry
Biggs:	Edwin Parker
Morse:	Ted Mapes
Sam Burger:	Sam Flint
Doctor:	Wheaton Chambers

DON COLORADO
#70

Director:	Derwin Abbe
Teleplay:	Sherman L. Lowe
Film Director:	Derwin Abbe
Cameraman:	Mack Stengler
Film Editor:	Fred Berger
Music Director:	Raoul Kraushaar
Copyright:	December 1, 1952
Season:	First, 1952-1953

Cast

Hopalong Cassidy:	William Boyd
Red Conners:	Edgar Buchanan
Hobart:	Nelson Leigh
Roger Endicott:	George Wallace
Noreen Thomas:	Noreen Nash
Doctor:	Stanley Blystone
Stage Driver:	Bud (Lennie) Osborne
Manager/Stage Agent:	John Frank

THE TRAP
#71

Director:	Derwin Abbe
Teleplay:	Sherman A. Lowe
Production Supervisor:	Glenn Cook
Cameraman:	Mack Stengler
Film Editor:	Fred Berger
Music Director:	Raoul Kraushaar
Composers:	Nacio Herb Brown and L. Wolfe Gilbert, *Hopalong Cassidy March*
Season:	First, 1952-1953
Copyright:	September 26, 1952

Cast

Hopalong Cassidy:	William Boyd
Red Conners:	Edgar Buchanan
Lou Forler:	Howard Negley
Tom Stacy:	Lane Bradford
Norman Blaine:	Bill Henry
Christine Russell:	Cajan Lee
Maudie:	Maudie Prickett

GHOST TRAILS
#72

Director:	Thomas Carr
Teleplay:	Harrison Jacobs
Production Supervisor:	Glenn Cook
Cameraman:	Mack Stengler
Film Editor:	Fred Berger
Music Director:	Raoul Kraushaar
Copyright:	October 28, 1952
Season:	First, 1952-1953

Cast

Hopalong Cassidy:	William Boyd
Red Conners:	Edgar Buchanan
Deacon Denby:	Frank Ferguson
Boyle:	Edward Clark
Gil:	Jack Harden
Steve:	Ted Mapes
Henchman:	John Cason
Morgan:	Tom London
Uncredited Bit Part:	Charles F. Seel
Uncredited Bit Part:	Frank Jaquet

GUNS ACROSS THE BORDER
#73

Director:	Thomas Carr
Teleplay:	Harrison Jacobs
Production Supervisor:	Glenn Cook
Camerman:	Mack Stengler
Music Director:	Raoul Kraushaar
Copyright:	September 19, 1952
Season:	First, 1952-1953

Cast

Hopalong Cassidy:	William Boyd

Red Conners:	Edgar Buchanan
Mamie Taylor:	Myra Marsh
Capt. Lee Sterling:	Keith Richards
Daly:	Henry Rowland

THE FEUD
#74

Director:	Thomas Carr
Teleplay:	Sherman L. Lowe
Production Supervisor:	Glenn Cook
Cameraman:	Mack Stengler
Music Director:	Raoul Kraushaar
Copyright:	October 28, 1952
Season:	First, 1952-1953

Cast

Hopalong Cassidy:	William Boyd
Red Conners:	Edgar Buchanan
Frank Scofield:	Hugh Beaumont
Nancy Croft:	Lucia Carroll
Neil Croft:	Steve Darrell
John Emery:	Harold Goodwin
Doctor:	Herbert Lytton

MARKED CARDS
#75

Director:	Derwin Abbe
Teleplay:	Harrison Jacobs
Production Supervisor:	Glenn Cook
Cameraman:	Mack Stengler
Film Editor:	Fred Berger
Music Director:	Raoul Kraushaar
Copyright:	October 28, 1952
Season:	First, 1952-1953

Cast

Hopalong Cassidy:	William Boyd
Red Conners:	Edgar Buchanan
Tommy Lewis:	Tommy Ivo
Brad Mason:	George Wallace
Sam Coates:	Crane Whitely
Jack Gardner:	James Diehl
Judge Graham:	Emmett Vogan
Dick Lewis:	John Deering

THE VANISHING HERD
#76

Director:	Derwin Abbe
Teleplay:	Harrison Jacobs
Production Supervisor:	Glenn Cook
Cameraman:	Mack Stengler
Film Editor:	Fred Berger
Music Director:	Raoul Kraushaar
Copyright:	December 12, 1952
Season:	First, 1952-1953

Cast

Hopalong Cassidy:	William Boyd
Red Conners:	Edgar Buchanan
Nelson Whitley:	Betty Ball
Col. Chavez:	Edward Colmans
Ramon Gardena:	Keith Richards
Dillon:	Lee Roberts
Burke:	Pierce Lyden
Bascomb:	Edgar Carpenter

BLACK WATERS
#77

Director:	Thomas Carr
Teleplay:	Harrison Jacobs
Production Supervisor:	Glenn Cook
Cameraman:	Mack Stengler
Film Editor:	Fred Berger
Music Director:	Raoul Kraushaar
Composers:	Nacio Herb Brown and L. Wolfe Gilbert, *Hopalong Cassidy March*
Copyright:	December 1, 1952
Season:	First, 1952-1953

Cast

Hopalong Cassidy:	William Boyd
Red Conners:	Edgar Buchanan
Johnny Tall Horse:	Rick Vallin
Chief Tall Horse:	Morris Ankrum
Betty Turner:	Marilyn Nash
Blaine Turner:	Walter Reed
Sheriff:	Clarence Straight
Brandon:	Malcom Beggs

BLIND ENCOUNTER
#78

Director:	Thopmas Carr
Teleplay:	Sherman L. Lowe
Copyright:	December 1, 1952
Season:	First, 1952-1953

Cast

Hopalong Cassidy:	William Boyd
Red Conners:	Edgar Buchanan
Manuel Soledad:	Pepe Hern
Senora Soledad:	Argentina Brunetti
Poynter:	Denver Pyle
Andrews:	Robert Bice
Carmencita Escobar:	Donna Martell
Corporal Gonzales:	John Halloran
Blackie:	Philip Van Zandt

THE PROMISED LAND
#79

Director:	Derwin Abbe
Teleplay:	Sherman L. Lowe
Production Supervisor:	Glenn Cook
Cameraman:	Mack Stengler
Film Editor:	Fred Berger
Music Director:	Raoul Kraushaar
Copyright:	December 10, 1952
Season:	First, 1952-1953

Cast

Hopalong Cassidy:	William Boyd
Red Conners:	Edgar Buchanan
Frank Dale:	John Crawford
Irene Mayo:	Maura Murphy
Arnold Rivers:	Thurston Hall
Slim:	Edwin Parker
Mr. Todd:	William Fawcett
Henchman/Shaw:	Sandy Sanders

LAWLESS LEGACY
#80

Director:	Derwin Abbe
Teleplay:	Royal Cole
Production Supervisor:	Glenn Cook
Cameraman:	Mack Stengler
Film Editor:	Fred Berger
Music Director:	Raoul Kraushaar
Composers:	Nacio Herb Brown and L. Wolfe Gilbert, *Hopalong Cassidy March*
Copyright:	December 31, 1952
Season:	First, 1952-1953

Cast

Hopalong Cassidy:	William Boyd
Red Conners:	Edgar Buchanan
Judy Marlow:	Claudia Barrett
Tom Marlow:	Stephen Rowland
Judd Marlow:	Tim Graham
Trimmer Lane:	Clayton Moore
Bob Hayes:	Marshall Reed
Sheriff:	Edgar Dearing

TRICKY FINGERS
#81

Director:	George Archainbaud
Teleplay:	Harrison Jacobs
Copyright:	April 2, 1954
Season:	Second, 1953-1954

Cast

Hopalong Cassidy:	William Boyd
Red Conners:	Edgar Buchanan

Adele Keller:	Marjorie Lord
Harry Keller:	Mark Dana
Albert Biggs:	Stanley Andrews

THE VALLEY RAIDERS
#82

Director:	George Archainbaud
Teleplay:	George W. Sayre
Copyright:	October 23, 1953
Season:	Second, 1953-1954

Cast

Hopalong Cassidy:	William Boyd
Red Conners:	Edgar Buchanan
Stephen Roberts:	Lyle Talbot
Morgan:	Henry Rowland
Doctor:	Harte Wayne

MASQUERADE FOR MATILDA
#83

Director:	Derwin Abbe
Teleplay:	Joe Richardson
Copyright:	March 5, 1954
Season:	Second, 1953-1954

Cast

Hopalong Cassidy:	William Boyd
Red Conners:	Edgar Buchanan
Matilda Harron:	Hazel Keener
Samuel Harron:	Phil Tead
Nail:	George Keymas
Jeb:	Zon Murray
Constable Tom Gorham:	Roy Barcroft
Breck:	Frank Marlowe

TWISTED TRAILS
#84

Director:	Derwin Abbe
Teleplay:	Joseph O'Donnell
Copyright:	October 30, 1953
Season:	Second, 1953-1954

Cast

Hopalong Cassidy:	William Boyd
Red Conners:	Edgar Buchanan
Frank Walton:	Herbert Lytton
Gale:	Richard Farmer
Pat Dougherty:	Gloria Talbott
Herb Dougherty:	George Spaulding
Buckley:	Lane Bradford
Cashier:	Wheaton Chambers
Bart:	Norman "Rusty" Wescoatt

DON'T BELIEVE IN GHOSTS
#85

Director:	George Archainbaud
Teleplay:	Ande Lamb
Copyright:	December 25, 1953
Season:	Second, 1953-1954

Cast

Hopalong Cassidy:	William Boyd
Red Conners:	Edgar Buchanan
Billy Murdock:	Anthony Sydes
Dorothy Murdock:	Aline Towne
Henry Grant:	Carleton Young
Frank Ellis:	Steve Pendleton
Miss Kerner:	Almira Sessions
Sheriff Lane:	Stanley Blystone

THE SOLE SURVIVOR
#86

Director:	George Archainbaud
Teleplay:	Ande Lamb
Production Supervisor:	Glenn Cook
Copyright:	October 16, 1953
Season:	Second, 1953-1954

Cast

Hopalong Cassidy:	William Boyd
Red Conners:	Edgar Buchanan
Fred Loomis:	Richard Reeves
Sheriff Gordon:	Kenneth MacDonald
Carol Madden:	Dorothy Green
Dick Madden:	David Bruce
John Henderson:	Harry Hines

FRAME-UP FOR MURDER
#87

Director:	Derwin Abbe
Teleplay:	Harrison Jacobs
Copyright:	March 12, 1954
Season:	Second, 1953-1954

Cast

Hopalong Cassidy:	William Boyd
Red Conners:	Edgar Buchanan
Lt. Rex Melton:	Bill Henry
Morse:	Ray Walker
Brenner:	Robert Knapp
Doctor:	Harry Hayden

ILLEGAL ENTRY
#88

Director:	George Archainbaud
Teleplay:	Harrison Jacobs
Copyright:	November 20, 1953
Season:	Second, 1953-1954

Cast

Hopalong Cassidy:	William Boyd
Red Conners:	Edgar Buchanan
Wilberforce Lawrence Edgemont/The Professor:	Emerson Treacy
Cal Foster:	Harry Lauter
Grimes:	Frank Hagney
Sing Lee:	Spencer Chan
Capt. Moreno:	Paul Marion
Jim Morgan:	Larry Hudson

THE BLACK SOMBRERO
#89

Director:	Derwin Abbe
Teleplay:	George W. Sayre
Copyright:	March 19, 1954
Season:	Second, 1953-1954

Cast

Hopalong Cassidy:	William Boyd
Red Conners:	Edgar Buchanan
Marco Rodriquez:	Rick Vallin
Dorothy Preston:	Cajan Lee
John Preston/Jose Hernandez:	Morris Ankrum
Belker:	Duane Thorsen
Carter:	Larry Hudson
Felix Sanchez:	Edward Colmans
Judge:	Forrest Taylor
Servant:	Tony Roux

THE LAST LAUGH
#90

Director:	Derwin Abbe
Teleplay:	Ande Lamb
Copyright:	November 6, 1953
Season:	Second, 1953-1954

Cast

Hopalong Cassidy:	William Boyd
Red Conners:	Edgar Buchanan
Mayor Hiram Tilden:	Edward Clark
Richard Wald:	Edgar Dearing
Don Moore:	Alan Wells
Sid Michael:	John Crawford
Betty Black:	Jeanne Dean

ARIZONA TROUBLE-SHOOTERS
#91

Director:	George Archainbaud
Teleplay:	William Lively
Copyright:	December 4, 1953
Season:	Second, 1953-1954

Cast

Hopalong Cassidy:	William Boyd
Red Conners:	Edgar Buchanan
Turk Shanns:	Richard Avonde
George Byers:	Mort Mills
Big John Bragg:	Howard Negley
Jack Lawson:	Gregg Barton
Jane Sawyer:	Nan Leslie
Frank Sawyer:	Ned Davenport

NEW MEXICO MANHUNT
#92

Director:	George Archainbaud
Teleplay:	William Lively
Copyright:	January 22, 1954
Season:	Second, 1953-1954

Cast

Hopalong Cassidy:	William Boyd
Red Conners:	Edgar Buchanan
Soakie:	Raymond Hatton
Frank Bent:	Russ Conway
Sam Hardin:	Leslie O'Pace
Stacy Keller:	Douglas Kennedy
Marian:	Dolores Mann
Dakota:	House Peters, Jr.

THE EMERALD SAINT
#93

Director:	George Archainbaud
Teleplay:	J. Benton Cheney
Copyright:	March 26, 1954
Season:	Second, 1953-1954

Cast

Hopalong Cassidy:	William Boyd
Red Conners:	Edgar Buchanan
Sam Chapman/Jim Forrester:	George Wallace
Don Miguel Alvarez:	Don Alvarado
Lola Alvarez:	Anna Navarro
Frazer:	Jack Ingram
Jose:	Salvador Baguez
Wilks:	Ted Bliss
Mexican Woman:	Julia Montoya

DOUBLE TROUBLE
#94

Director:	Derwin Abbe
Teleplay:	Joe Richardson
Copyright:	January 8, 1954
Season:	Second, 1953-1954

Cast

Hopalong Cassidy:	William Boyd
Red Conners:	Edgar Buchanan
Jose Martinez:	Victor Millan
Maria Martinez:	Charlita Roeder
Ollie Gray:	Sam Flint
Ted Gray:	Robert Knapp
Trig Dawson:	John Pickard
Monte Kane:	Donald Novis

THE JINX WAGON
#95

Director:	George Archainbaud
Teleplay:	Robert Schaefer/Eric Freiwald
Copyright:	November 13, 1953
Season:	Second, 1953-1954

Cast

Hopalong Cassidy:	William Boyd
Red Conners:	Edgar Buchanan
Dan Clemens:	Thurston Hall
Ginny Clemens:	Kathleen Case
Jeff Clemens:	Michael Thomas
Lait:	Steve Conte
Rick Bayless:	Myron Healey
Sandy Morgan:	Paul Burns

SILENT TESTIMONY
#96

Director:	George Archainbaud
Teleplay:	Ande Lamb
Copyright:	February 19, 1954
Season:	Second, 1953-1954

Cast

Hopalong Cassidy:	William Boyd
Red Conners:	Edgar Buchanan
Pop Ashton:	Hank Patterson
Rick Ashton:	James Best
Trixie Brown:	Vici Raaf
Jason Rand:	Richard Cutting
Pike Jenning:	Keith Richards
Marshal:	Steve Clark

DEATH BY PROXY
#97

Director:	Derwin Abbe
Teleplay:	Cecile Kramer
Copyright:	December 11, 1953
Season:	Second, 1953-1954

Cast

Hopalong Cassidy:	William Boyd
Red Conners:	Edgar Buchanan
Jim Adams:	Paul Richards
Doc Weston:	Fred Sherman
Slade:	Duane Thorsen (Walter Reed)
Burke:	Pierce Lyden (Wesley Hudman)
Frank Samuels:	Charles Cane
Bank Clerk:	John Deering

THE RENEGADE PRESS
#98

Director:	George Archainbaud
Teleplay:	Robert Schaefer / Eric Freiwald
Copyright:	January 1, 1954
Season:	Second, 1953-1954

Cast

Hopalong Cassidy:	William Boyd
Red Conners:	Edgar Buchanan
Ramon Torres:	Rick Vallin
Tom McLean:	William Fawcett
Steve:	Terry Frost
Frank Harrison:	William Phillips
Jim Woods:	Lou Nova
Wade:	I. Stanford Jolley

THE DEVIL'S IDOL
#99

Director:	George Archainbaud
Teleplay:	Robert Schaefer / Eric Freiwald
Copyright:	October 9, 1953
Season:	Second, 1953-1954

Cast

Hopalong Cassidy:	William Boyd
Red Conners:	Edgar Buchanan
Johnny Bolton:	Ron Hagerthy
Rev. Edward Adams:	Nolan Leary
Burke Ramsey:	Don C. Harvey
Ed Garver:	Harry Harvey
Ralph Higgins:	Danny Mummert

GYPSY DESTINY
#100

Director:	Derwin Abbe
Teleplay:	Ande Lamb
Copyright:	November 27, 1953
Season:	Second, 1953-1954

Cast

Hopalong Cassidy:	William Boyd
Red Conners:	Edgar Buchanan
Artaro:	Robert Cabal
Marella:	Pilar del Rey
Mother Kiomi:	Belle Mitchell
Jeff:	Paul Richards
Charley Mitchell:	John Merton
King Lasho:	Frank Lackteen

3-7-77
#101

Director:	George Archainbaud
Teleplay:	Ande Lamb
Copyright:	February 26, 1954
Season:	Second, 1953-1954

Cast

Hopalong Cassidy:	William Boyd
Red Conners:	Edgar Buchanan
Sheriff Ray Barnett:	James Anderson
David Barnett:	James Seay
Chad Young:	Ted Stanhope
Sam Wade:	Leonard Penn
Earl Fox:	Dick Rich
Stage Driver:	Bud Osborne

COPPER HILLS
#102

Director:	Derwin Abbe
Teleplay:	Harrison Jacobs
Copyright:	January 15, 1954
Season:	Second, 1953-1954

Cast

Hopalong Cassidy:	William Boyd
Red Conners:	Edgar Buchanan
Tommy Red Arrow:	Joseph Waring
Judson Rush:	George Wallace
Constable Martin:	Edwin Rand
Leeds:	Lee Roberts
Picture Pete:	Earle Hodgins
Nat Burnham:	Paul Birch

THE OUTLAW'S REWARD
#103

Director:	George Archainbaud
Teleplay:	Robert Schaefer / Eric Freiwald
Copyright:	January 29, 1954
Season:	Second, 1953-1954

Cast

Hopalong Cassidy:	William Boyd
Red Conners:	Edgar Buchanan
Dr. Glenn Scott:	Harlan Warde
Charles Scott:	Griff Barnett
Frank Prescott:	John Alvin
Nancy Mathews:	Elaine Riley
Vic James:	Denver Pyle
Al:	William Haade

FRONTIER LAW
#104

Director:	Derwin Abbe
Teleplay:	William Lively
Copyright:	December 18, 1953
Season:	Second, 1953-1954

Cast

Hopalong Cassidy:	William Boyd
Red Conners:	Edgar Buchanan
Clay Morgan:	Bill Henry
Buck Staley:	Robert Griffin
Jenny Warren:	Barbara Knudson
Henry Warren:	Pierre Watkin

Rawhide Carney:	Marshall Reed
Bearcat Smith:	Daniel White

GRUBSTAKE
#105

Director:	Derwin Abbe
Teleplay:	Ande Lamb
Copyright:	February 2, 1954
Season:	Second, 1953-1954

Cast

Hopalong Cassidy:	William Boyd
Red Conners:	Edgar Buchanan
Dr. Sheldon Lowe:	Christopher Dark
Bummer Lowe:	Percy Helton
Mrs. Turner:	Gladys George
Trem Hardy:	Robert Paquin
Dan Warner:	Timothy Carey
Brock Fane:	Michael Fox

STEEL TRAILS WEST
#106

Director:	Derwin Abbe
Teleplay:	Harrison Jacobs
Copyright:	February 12, 1954
Season:	Second, 1953-1954

Cast

Hopalong Cassidy:	William Boyd
Red Conners:	Edgar Buchanan
Bill Bricker:	Richard Powers
Sam Murdock:	Robert Bice
Jim Taggart:	Lewis Martin
Bob Murdock:	Donald Kennedy

TRAILERS

TRAILER #1
LENGTH: 36 feet
PRODUCTION   #81 (Tricky Fingers)
                    #90 (The Last Laugh)

FADE IN
SETTING: Interior office, medium close shot, daytime. Hoppy seated at desk, glancing through some papers, looks directly into camera, smiles, and says...
TEXT: Hi, little partners...Have you been doing anything to help Mom around the house lately? You haven't? Let's do all we can to help her, eh? I'll bet you have fun doing it - and I know Mom will appreciate it. Would you do that for me? Now, 'til next week, so long - and good luck! (Hoppy makes gesture of farewell, smiles...)
FADE OUT

TRAILER #2
LENGTH: 45 feet
PRODUCTION #82 (The Valley Raiders)
                    #104 (Frontier Law)
FADE IN
SETTING: Interior Hoppy's office, medium close shot. Hoppy seated at desk reading papers, glances into camera, smiles, puts down papers and speaks...
TEXT: Hi, there...did you hear what the man said? I did - and I agree with every word of it. If you buy his product I'm sure you'll like it - and it'll tickle my sponsor to death. Now a thought for my little partners...be happy and proud that you are an American. Remember, there are millions of little boys and girls all over the world that would give anything for that privilege. Until next week - so long and good luck to you. (Hoppy smiles and waves...)
FADE OUT

TRAILER #3
LENGTH: 54 feet
PRODUCTION   #83 (Masquerade For Matilda)
                    #88 (Illegal Entry)

FADE IN
SETTING: Interior Hoppy's office, close up shot, daytime. Hoppy looking into camera and talking...
TEXT: Hi, there...you know, there are certain laws that are made to protect all of us - but we must do something to help about that protection. And here's a thought that might help prevent an accident. Help other people protect you. Now to make it a little easier for you to remember those words...the first letter of each of those words spell the name of a man who thinks you're pretty wonderful, and who doesn't want you to be hurt. Until next week - so long. In the meantime, be careful, won't you? (Hoppy smiles...)
FADE OUT

TRAILER #4
LENGTH: 41 feet
PRODUCTION   #84 (Twisted Trails)
                    #85 (Don't Believe In Ghosts)
FADE IN
SETTING: Interior Hoppy's office, medium close shot. Hoppy seated behind his desk, papers in hand, glances up, suddenly realizes he is facing camera, smiles and says directly into camera...
TEXT: Hi, there, my little friends. Have you been washing your face and hands lately? How about your ears? Oh, that's pretty good...don't forget that soap and water not only keeps you nice and clean, but it also protects your health, so use lots of it, won't you? I'll see you next week. (Hoppy smiles and waves his hand...)
FADE OUT

TRAILER #5
LENGTH: 53 feet
PRODUCTION #86 (The Sole Survivor)
                    #101 (3 - 7 - 77)
FADE IN
SETTING: Interior Hoppy's office, close up shot on Hoppy looking directly into camera as he says...
TEXT: Hi, there....friends. This is the first time in the five years since you've been kind enough to invite me into your homes on television, that I've taken the liberty of speaking to you personally. I would just like to assure you that any product you see Hoppy's name on, in my mind, is the finest. Now here's a little message for my little friends...always mind your Daddy and your Mommy. Please remember, they're the best friends you have in the world. Until next week - take good care of yourselves. (Hoppy steps back with a smile and a wave of his hand...)
FADE OUT

TRAILER #6
LENGTH: 43 feet
PRODUCTION   #87 (Frameup For Murder)
                    #95 (The Jinx Wagon)
FADE IN
SETTING: Interior Hoppy's office, medium close shot, daytime. Hoppy seated behind desk, looks up, smiles and says...
TEXT: Hi, there, little friends. This is just a thought about guns. I was doing a scene one time, in one of the pictures,

and the boy that was working with me was twirling his gun in his hand and playing with it. I said, "What are you doing with that gun, Lucky?" He said, "I'm just playing with it!" I said, "Those things are not to play with...they kill people!" So watch your guns, children. Be careful with them, won't you?...
FADE OUT

TRAILER #7
LENGTH: 54 feet
PRODUCTION #89 (The Black Sombrero) #91 (Arizona Troubleshooters)
FADE IN
SETTING: Interior Hoppy's office, medium close shot. Hoppy, glancing up from book, says...
TEXT: Hi, my young friends. You know, I've had a lot of letters in the last six or eight months about you not wearing your raincoats - and your overshoes. Now the reason you don't see 'Hoppy' in a raincoat and overshoes, is because I live in a country where there isn't too much rain...but you **must - always** - whenever it's raining, wear your raincoats and your overshoes. And there's one more thought, too - don't leave them at the schoolhouse - bring them home, will you? (Hoppy chuckles) I'll see you next week, and be sure and wear them, will you?...
FADE OUT

TRAILER #8
LENGTH: 33 feet
PRODUCTION #92 (New Mexico Manhunt) #106 (Steel Trails West)
FADE IN
SETTING: Exterior country, under a tree, medium close shot, daytime. Hoppy mounted on his horse, waves as he looks into camera and says...
TEXT: Hi, little partners. Any time you play a game - play it to win. But remember, somebody always has to lose. And in case it's you, don't be unhappy about it. Be a good loser - just try a little harder the next time. I'll see you next week. (Hoppy waves and smiles...)
FADE OUT

TRAILER #9
LENGTH: 55 feet
PRODUCTION #93 (The Emerald Saint) #99 (The Devil's Idol)
FADE IN
SETTING: Exterior country, medium close shot on Hoppy, daytime. Building in background, Hoppy stands near the fence near his horse, Topper...
TEXT: Hi, there, little partners. I think you know your Mommy is the nicest and the most beautiful woman in the world - and that she loves you very dearly. So when she asks you to do anything, don't fuss about it - just do it.

For instance - when she asks you to have a glass of milk - drink it - and then surprise her by asking for another one. You try that - and see how much better you both feel. Now until next week, so long - and in the meantime - watch yourselves at the crosswalks, will you?...
FADE OUT

TRAILER #10
LENGTH: 91 feet
PRODUCTION #94 (Double Trouble)
FADE IN
SETTING: Interior sheriff's office, medium close shot, daytime. Hoppy seated on edge of table, looking into camera...
TEXT: Hi, my little partners. There's a group of men in America who walk the streets, and stand on the corners, dressed in pretty good-looking uniforms. Those men look pretty smart - and I want to assure you that they are smart, or they wouldn't be there. I'm speaking of policemen! The policeman is a father - the same as some of you grownups are. He has children the same as you have. And his job is to help you protect your children - and to help protect the children themselves. So keep one thing in mind, won't you, please? Don't call him a 'cop'. He's a respected man, and we should call him by a respected name - a 'police officer'. Always remember that, and see how much nicer he'll be. I'll see you next week. (Hoppy waves to his audience...)
FADE OUT

TRAILER #11
LENGTH: 63 feet
PRODUCTION #96 (Silent Testimony) #97 (Death By Proxy)
FADE IN
SETTING: Interior living room, medium close shot, daytime. Hoppy seated, reading book, closes book as he looks into camera and says...
TEXT: Hi, there, little partners. I understand there's been quite a little talk around your house lately about 'going to bed'. I know that we all hate to go to bed. We're always so afraid we're going to miss something. But you must get a lot of sleep, and the time to go to bed is when your Mommy or your Daddy tell you to - not fifteen minutes later, or twenty minutes later, or one or two television shows later - but right now. I'll see you next week. Until then, get lots of good sleep, will you? Good night. (Hoppy waves and picks up book...)
FADE OUT

TRAILER #12
LENGTH: 71 feet
PRODUCTION #98 (Renegade Press) #103 (The Outlaw's Reward)
FADE IN

SETTING: Interior sheriff's office, medium close shot, daytime. Hoppy leaning on table, looks into camera as he waves a greeting...
TEXT: Hi, my little partners. I want to ask you a very serious question. Are you afraid of policemen? You're not, huh? Well...that's good! You shouldn't be! Always remember that those men have had a lot of training, and put in a lot of years to learn how to help protect you. And one more thing - don't ever call a policeman a 'cop'...he doesn't like it, and it doesn't sound good. Call him a police officer, and see how much nicer he'll be. I'll see you next week - and in the meantime, don't forget to go to Sunday School...
FADE OUT

TRAILER #13
LENGTH: 47 feet
PRODUCTION #100 (Gypsy Destiny)
FADE IN
SETTING: Interior room, close up shot, daytime. Hoppy, smiles into camera...
TEXT: Hi, there, friends. You know sometimes a smile will buy things that money can't buy. But there's something pretty important that goes with that smile...good teeth! A lot of people worry about going to the dentist...remember - he's just as nice a guy as any other guy, and he wants to help you. So go and see him...at least once a year - maybe twice a year. Brush your teeth good twice every day, too...will you do that for me? (Hoppy grins into camera...)
FADE OUT

TRAILER #14
LENGTH: 63 feet
PRODUCTION #102 (Copper Hills) #105 (Grubstake)
FADE IN
SETTING: Interior Hoppy's office, medium close shot, daytime. Sheriff's bulletin board behind Hoppy. Hoppy looks directly into camera...
TEXT: Hi, there, my little friends. If you were playing baseball, and the baseball was knocked across a railroad track, I'm pretty sure that when you got to that railroad track you'd stop - look - and listen, wouldn't you? Now please remember that any street in America is just as dangerous as that railroad - so when you get to the curbstone, stop - look - and listen! Will you do that for me? Now so long until next week - and in the meantime - don't forget to go to Sunday School. (Hoppy grins, gestures farewell...)
FADE OUT

TELEVISION STATIONS

The Hopalong Cassidy TV program appeared on the following major television stations during the spring of 1953:

Albuquerque	KOB
Atlanta	WSB-TV
Baltimore	WBAL-TV
Binghamton	WNBF
Boston	WBZ
Buffalo	WBEN-TV
Charlotte	WBTV
Chicago	WNBQ
Cincinnati	WLW-T
Cleveland	WNBK
Columbus	WLW-C
Dallas	WFAA
Dayton	WLW-D
Denver	KFEL
Detroit	WWJ-TV
Erie	WICU
Grand Rapids	WOOD-TV
Greensboro	WFMY-TV
Honolulu	KGMB
Houston	WPRC-TV
Jacksonville	WMBR-TV
Johnstown	WJAC-TV
Kansas City	WDAF-TV
Los Angeles	KNBH
Louisville	WAVE-TV
Memphis	WMCT
Miami	WTVJ
Milwaukee	WTMJ-TV
Mobile	WALA
Nashville	WSM
New Haven	WNHC-TV
New Orleans	WDSU-TV
New York	WNBT
Norfolk	WTAR-TV
Oklahoma City	WKY-TV
Omaha	WOW-TV
Philadelphia	WPTZ
Phoenix	KPHO-TV
Pittsburgh	WDTV
Portland	UHF
Providence	WJAR-TV
Reading	WHUM
Roanoke	WROV
Rochester	WHAM-TV
St. Louis	KSD-TV
St Paul-Minneapolis	WSTP-TV
Salt Lake City	KDYL-TV
San Antonio	KEYL
San Francisco	KRON-TV
Schenectady	WRGB
Seattly	KING-TV
Spokane	UHF
Syracuse	WSYR-TV
Toledo	WSPD-TV
Toronto	CBC-TV
Tulsa	KOTV
Washington	WNBW

RADIO PROGRAMS AND STATION LISTINGS

Radio Shows

1. Dead Man's Hand
2. Rainmaker of Eagle Nest Mountain
3. Coltsville Territory
4. Mystery of Skull Valley
5. Renegade of San Rafael
6. Phantom Bandido
7. Murder on the Trail
8. Hoppy Takes a Chance
9. Voice of the Dead
10. Ten Strike Gold
11. Red Rock Mesa
12. Empty Saddle
13. The Failure
14. Bandits of Ridge Creek
15. Killers of Sandy Gulch
16. Red Death
17. Coyote's Creed
18. Bullets For Ballots
19. Green Valley Payoff
20. Man Who Made Willy Whirl
21. Range War
22. Letter from the Grave
23. Death Paints a Picture
24. Border of Nowhere
25. Medicine Man
26. Flying Outlaw
27. Sundown Kid
28. Hoppy Sees Red
29. Hoppy and the Schoolmarm
30. The King of Cinnabar
31. The Shell Game
32. Blood Money
33. The Disappearing Deputy
34. The Whistling Ghosts
35. An Old Spanish Custom
36. Secret of Martin Doon
37. Four to Go
38. The Red Terror
39. Hoppy and the Iron Horse
40. Gunsmoke Rides the Stagecoach Trail
41. Tinker's Dam
42. Hoppy Settles a Debt
43. Hoppy Turns on the Heat
44. Death Runs Dry
45. Melody of Murder
46. The King of Spades
47. Hoppy Takes the Bull by the Horns
48. Dead or Alive
49. Mystery at Tree Oaks
50. Plague of Parsons Folly
51. Mystery at the Diamond
52. The Golden Lure
53. Case of the Last Word
54. Bad Medicine at Rimrock
55. The Frightened Town
56. The Killer in Black
57. Coming Attraction Murder
58. Wet Beef and Dry Bones
59. Black Grass Fever
60. The Cold Country
61. Buckshot Badman
62. Boss of Vinegar Bend
63. Land of the Gunhawks
64. The Devil's Drum
65. Hoppy Elects a Sheriff
66. Peril at Pier 19
67. Death Crosses the River
68. Stagecoach West
69. The Unwilling Outlaw
70. Kidnapper's Trail
71. Bandit of Blackton Bend
72. Hook, Line and Murder
73. The Phantom Panther
74. Hoppy Plays a Hunch
75. A Jailer Named Satan
76. Man in the Yellow Mask
77. Run, Sheep, Run
78. Hoppy Meets His Match
79. Apaches Don't Need Guns
80. A Shot in the Dark
81. Gunhawk Convention
82. Gunfighter in Short Pants
83. Songbird of Santa Fe
84. Bayou Drums
85. Cleanup of Caribou Mesa
86. Six Little Men Who Were Green
87. Junior Badman
88. Devil and El Diablo
89. Lawyer of Laredo
90. Secret in the Hill
91. Memory of Mace Melot
92. Spider Woman
93. Killers of Lion Canyon
94. The Wastrels of Juarez
95. Gambler's Luck
96. Danger Wears Two Faces
97. California or Bust
98. Death Comes Invited
99. Bullfight
100. Women of Windy Ridge
101. Right Rope - Wrong Neck
102. Stampede at Semple Crossing
103. Cowtown Troubleshooters
104. Santa Claus Rustlers

Radio Stations

The Hopalong Cassidy program was featured on these major radio stations during the spring of 1953:

Abilene, TX	KRBC
Albany, OR	KWIL
Amarillo, TX	KLYN
Augusta, GA	WJBF
Augusta, ME	WRDO
Billings, MT	KOOK
Birmingham, AL	WAPI
Bristol, CT	WBIS
Bristol, TN	WOPI
Burlington, IA	KBUR
Butte, MT	KBOW
Charleston, SC	WTMA
Decatur, AL	WHOS
Dodge City, KS	KGNO
Durham, NC	WTIK
East Liverpool, OH	WOHI
El Paso, TX	KSET
Enid, OK	KCRC
Eugene, OR	KASH
Fresno, CA	KFRE
Gainesville, GA	WGGA
Georgetown, SC	WGTN
Great Bend, KS	KVGB
Gulfport, MS	WGCM
Hutchinson, KS	KWBW
Iowa City, IA	KXIC
Jamestown, NY	WJTN
Johnson City, TN	WETB
Kalamazoo, MI	WKZO
Kennewick, WA	KWIE
Kinston, NC	WFTC
Kirksville, MO	KIRX
Las Vegas, NV	KENO
Lincoln, NE	KFOR
Little Rock, AK	KVLC
Lubbock, TX	KFYO
Madisonville, KY	WFMW
Mobile, AL	WKRG
Oskaloosa, IA	WBOG
Ottumwa, IA	KBIZ
Phoenix, AR	KTAR
Prairie Du Chien, WI	WPRE
Presque Isle, ME	WAGM
Rapid City, SD	KOTA
Roanoke, VA	WSLS
Sacramento, CA	KROY
San Angelo, TX	KGKL
Sikeston, MO	KSIM
Sioux Falls, SD	KISD
Tallahassee, FL	WTNT
Tuscaloosa, AL	WJRD
Tucson, AR	KOPO
Walterboro, SC	WALD
Wichita, KS	KANS
Wilkes Barre, PA	WILK
Yakima, WA	KIMA
Yuma, AR	KYUM

Honolulu
Luxembourg
Canada

Appendix 5
COMIC STRIP LISTINGS

In 1953, the Hopalong Cassidy comic strip was a regular feature of the following newspapers:

UNITED STATES - Daily

Abilene, TX *Reporter News*
Albany, NY *Times-Union*
Allentown, PA *Call*
Atlanta, GA *Metropolitan Herald*
Baltimore, MD *News-Post*
Bangor, ME *Daily Commercial*
Boise, ID *Statesman*
Boston, MA *Record American*
Cedar Rapids, IA *Gazette*
Charlotte, NC *Observer*
Chicago, IL *Herald-American*
Dallas, TX *News*
Detroit, MI *Times*
Du Bois, PA *Courier Express*
Dodge City, KS *Globe*
Dunn, NC *Daily Record*
Emporia, KS *Gazette*
Endicott, NY *Bulletin*
Galesburg, IL *Register Mail*
Green Bay, WI *Press Gazette*
Greenville, PA *Record Argus*
Groveille, CA *Mercury Register*
Hartford, CT *Times*
Huntington, WV *Herald Dispatch*
Jamestown, NY *Sun*
Kansas City, KS *Kansan*
Kansas City, MO *News Press*
Kingsport, TN *Times & News*
Knoxville, TN *Journal*
Lorain, OH *Journal*
Los Angeles, CA *Herald-Express*
Lowell, MA *Sun*
Lubbock, TX *Avalanche Journal*
Manchester, NH *Union Leader*
Mansfield, OH, *News Journal*
Meriden, CT *Star*
Miami, FL *News*
Nashville, TN *Tennessean*
New Albany, IN *Tribune*
Newanne, IL *Star Courier*
New Bedford, MA *Standard Times*
New Orleans, LA *Item*
New York, NY *News*
Norfolk, VA *Ledger Dispatch*
Ocalo, FL *Star Banner*
Oneonta, NY *Star*
Orlando, FL *Sentinel Star*
Oskaloosa, IA *Herald*
Oxnard, CA *Press Courier*
Pasco, WA *Tri-City Herald*
Pekin, IL *Times*
Peoria, IL *Journal Star*
Philadelphia, PA *Bulletin*

Pittsburgh, PA *Sun-Telegram*
Portland, IN *Commercial Review & Sun*
St. Paul, MN *Pioneer Press*
St. Petersburg, FL *Times*
Salem, OR *Capital Journal*
Salt Lake City, UT *Tribune Telegram*
San Antonio, TX *Light*
San Diego, CA *Union & Tribune Sun*
San Francisco, CA *Call-Bulletin*
Seattle, WA *Post-Intelligencer*
Springfield, MO *News Leader & Press*
Stroudsburg, PA *Record*
Syracuse, NY *Post Standard*
Tampa, FL *Times*
Texarkana, TX *Gazette & News*
Thomasville, GA *Times Enterprise*
Tifton, GA *Daily Gazette*
Tiffin, OH *Advertiser Tribune*
Toledo, OH *Times*
Van Wert, OH *Times Bulletin*
Washington, DC *Times Herald*
Waterbury, CT *Republican*
Wichita, KS *Beacon*
Worcester, MA *Telegram*
Youngstown, OH *Vindicator*

UNITED STATES - Sunday

Albany, NY *Times-Union*
Baltimore, MD *American*
Bluefield, WV *Telegraph & Sunset News*
Boise, ID *Statesman*
Boston, MA *Advertiser*
Cedar Rapids, IA *Gazette*
Charlotte, NC *Observer*
Chicago, IL *Herald-American*
Dallas, TX *News*
Detroit, MI *Times*
Dodge City, KS *Daily Blade*
Greenville, NC *Reflector*
Joplin, MO *Globe & News Herald*
Knoxville, TN *Journal*
Los Angeles, CA *Examiner*
Lowell, MA *Sun*
Miami, FL *News*
Nashville, TN *Tennessean*
New Albany, IN *Tribune*
New Bedford, MA *Standard Times*
New Orleans, LA *Item*
New York, NY *News*
Norfolk, VA *Ledger Dispatch*
Orlando, FL *Sentinel Star*
Peoria, IL *Journal Star*
Philadelphia, PA *Bulletin*
Pittsburgh, PA *Sun-Telegram*
Portland, OR *Journal*
St. Paul, MN *Dispatch & Pioneer Press*
St. Petersburg, FL *Sentinel Star*
Salem, OR *Capital Journal*
Salt Lake City, UT *Telegram*

San Antonio, TX *Light*
San Diego, CA *Union & Tribune Sun*
San Francisco, CA *Examiner*
Seattle, WA *Post-Intelligencer*
Springfield, MA *News Leader & Press*
Syracuse, NY *Herald Journal*
Toledo, OH *Blade*
Washington, DC *Times Herald*
Waterbury, CT *Republican*
Wichita, KS *Beacon*
Worcester, KS *Telegram*
Youngstown, OH *Indicator*

FOREIGN - Daily

Adelaide, Australia *News*
Barranquilla, Colombia *El Nacional*
Borgen, Norway *Morgensvisen*
Brussels, Belgium *Week-End*
Buenos Aires, Argentina *Clarin*
Ciudad Trujillo, Dominican Republic *El Nacion*
Copenhagen, Denmark *Edstrabladet*
Dublin, Ireland *Irish Independent*
Edinborough, Scotland *Edinborough Dispatch*
Edmonton, Canada *Journal*
Ft. Williams, Ontario, Canada *Daily Times Journal*
Frederiction, Canada, *Daily Gleanor*
Gothenburg, Sweden *N.Y. Tid.*
Guayquil, Ecuador *La Nacion*
Halifax, England *Halifax Courier*
Halmstad, Sweden, *Halland*
Hamilton, Bermuda *Mid-Ocean News*
Hamilton, Ontario, Canada *Spectator*
Havana, Cuba *El Pais*
Helsinki, Finland *Uusi Suomi*
Hongkong, China *Sunday Post Herald*
Honolulu, Hawaii, *Star Bulletin*
Hudikvalla, Sweden *Hudikvallis Tidningen*
Karlskrona, Sweden, *Sydoestra Soveriges*
Karlstad, Sweden *Vatermlands Folkblad*
Kristlandsand, Norway *Soerlandet*
Linkoeping, Sweden *Ostsvea Correspondenten*
Malmo, Sweden *Svallposten*
Manila, Philippine Islands *Philippines Herald*
Maracaibo, Venezuela *Informaciones*
Melbourne, Australia *Sun News Pictorial*
Mexico City, Mexico *Daily News*
Monterrey, Mexico *El Sol*
Montevideo, Uruguay *La Tribuna*
Montreal, Canada *Patrie*
Oslo, Norway *Aftonposten*
Ottawa, Canada *Citizen*
Porsgrunn, Norway *Porsgrunn Dagblad*

Randers, Denmark *Randers Antsavis*
Rio de Janeiro, Brazil *Empressa Journalistira Braseleira*
St. John, New Brunswick, Canada *Telegraph Journal*
San Juan, Puerto Rico, *El Imparcial*
San Jose, Costa Rica *Republica*
San Luis Potosi, Mexico *El Heraldo*
San Salvador, El Salvador *Prensa Grafica*
Stavenger, Norway *Stavenger Aftenblad*
Stockholm, Sweden *Empressen*
Sydney, Australia *Sun*
Toronto, Ontario, Canada *Star*
Trondheim, Norway *Abeider Avisen*
Vancouver, British Columbia, Canada *Star*
Victoria, Canada *Daily Times*
Windsor, Ontario, *Canada Star*

FOREIGN - Sunday

Aalborg, Denmark *Aalborg, Stifetidende*
Adelaide, Australia *Mail*
Agama, Guam *Guam News*
Bogota, Colombia *Diario Grafico*
Caraxas, Venezuela *El National*
Colombo, Ceylon *Ceylon Observer*
Copenhagen, Denmark *Carl Allers Establissement*
Haeraesand, Sweden *Vaestermorelands Allehand*
Hamilton, Bermuda *Mid-Ocean News*
Havana, Cuba *El Pais*
Honolulu, Hawaii *Star Bulletin*
London, Ontario, Canada *Free-Press*
Madras, India *Express Newspapers Ltd.*
Manila, Philippine Islands *Philippines Herald*
Melbourne, Australia *Sun News Pictorial*
Mexico City, Mexico *El Sol*
Montreal, Canada *La Patrie*
Montreal, Canada *Standard*
Muenchen, Germany *Sontagspost*
Oslo, Norway *Barnas, Avis*
Ottawa, Canada *Citizen*
Panama City, Panama *American*
Perth, Australia *Western Mail*
St. John, New Brunswick, Canada *Telegraph Journal*
San Juan, Puerto Rico *Rico El Emparcial*
San Salvador, El Salvador *La Prensa Grafica*
Stockholm, Sweden *Arst. Sun*
Stockholm, Sweden *Dagens Nyheter*
Sydney, Australia *Sunday Sun*
Toronto, Ontario, Canada *Star*
Tribune, Australia *Sunday Mail*
Vancouver, British Columbia *Sun*

PRICE GUIDE TO HOPALONG CASSIDY LICENSED MERCHANDISE

A price guide is only a guide. It is designed to get you into the ballpark and up to the plate. How you bat is entirely up to you. **It is not a price absolute**. Using it that way is an invitation to disaster.

In the 1990s market, value is influenced primarily by three key factors - condition, scarcity, and desirability. There are other factors, like affordability, that you will need to weigh, but these are the three most important. Take them all into consideration when buying Hopalong Cassidy items.

While I strongly believe that serious Hopalong Cassidy collectors should not buy items in less than C7 (fine) condition on a scale of C1 to C10, the truth is that the vast majority of Hoppy items that survive are at grade C6 (very good) condition or below. A C6 object is one that meets the following three criteria: (1) lovingly played with, (2) complete, and (3) no damage visible at arm's length. Less than one-third, perhaps even less than one quarter, of the Hopalong Cassidy items that survive meet this test. Many fall into the C5 (good) grade, i.e., complete but heavily played with and with damage to the visual surface evident at arm's length. An object that is incomplete, no matter how little is missing, must be graded at C4 (fair) or less.

The values used in this book are for objects in C6 condition. Fix this in your mind. Because they refer to objects that are complete and show no damage except upon close examination, these values apply to less than one-third the Hoppy objects that you will encounter. They are values for collectors, those who are willingly to pay the price for objects that they view as having long term collectible, and possibly investment, potential.

The biggest mistake made by Hopalong Cassidy collectors is underestimating the number of examples that survive for any given object. Because I have done research among the royalty statements in the William Boyd papers, I think in terms of hundreds of thousands and millions. I know only a few collectors and dealers that think as I do. Dealers are optimists. They want to believe that every object they find is scarce. The vast majority of what they find is extremely common. No Hoppy item is as scarce as dealers would have you believe.

What this means is that it pays to comparison shop. It is common to find the price difference between two dealers for the same item in the same condition differing by several hundred percent. In the mid-1990s, the general attitude among dealers is "price the object to its limits."

There is little question in my mind that the prices for Hopalong Cassidy licensed products in the mid-1990s are usurious. I buy actively, and I know what I pay. I pay far less than the published prices in numerous price guides that list Hopalong Cassidy objects. A few authors, and Ted Hake is **not** one of them, are quite willing to report exorbitant dealer pricing. It is totally self-serving. If they are dealers, they justify what they have charged their customers by showing them that the price appears in a printed source. If they are collectors, they enhance the value of what they own. One thing is certain, they did not pay these prices.

Do buyers occasionally pay some of these exaggerated prices? Tragically, the answer is often yes. The old adage "fools and their money are soon parted" applies. This is especially true when a Hoppy item appears at an auction. Auction fever dominates; auction regret comes later. However, one fool does not a market make.

The prices in this book are what an experienced Hoppy collector, one with a thorough knowledge of the merchandise and the market, will pay. You can buy at these prices if you are patient and persist in the hunt. Remember, you set value by what you pay, not the seller by what he asks. Vote with your wallet. If you have no self-control when you whip it out, you are at fault. Do not let the dealer intimidate you. When you do not like the price, walk away. Not every seller will get the message. Some individual's arrogance and the appearance of an occasional fool prevents this from happening. However, a few dealers will respond to the wake up call. These are the ones with whom you want to do business.

Do not confuse desirability with availability. Some common items, like the Hopalong Cassidy lunch kit, are very desirable and readily available. Other items, like Hopalong Cassidy Bell blue jeans, are scarce and not particularly desirable. The concept of desirability as a pricing factor is relatively new among collectors. It is the most subjective of the three - condition, scarcity, and desirability. Further, it is a concept that continually changes, driven in part by crossover collectors. In determining the prices that appear below, I have viewed desirability from the point of view of a Hoppy, not a crossover, collector. As a result, the Hopalong Cassidy radio is valued at $350, not the $1,000 plus that dealers ask in the figural radio market, a market which, by the way, is collapsing.

The values below are for objects without their original packaging. How much does the box add to the value? The answer begins at 50% and ends at 400%. The key is box's pizzazz coupled with the image on and availability of the box. You will have to judge the added value for yourself. In a few instances, I have provided the box value or box added value in the price listings to provide comparables.

The price listing has been kept simple. Detailed descriptions are used only when an object has not been illustrated previously or is necessary to differentiate between two similar objects. Occasionally dimensions are given. If you want to know the manufacturer of the item, check Appendix #1.

When necessary objects like bubble gum cards, milk bottles, etc., have been grouped and unit prices assigned. This is a price guide, not an encyclopedic checklist of every licensed Hoppy object.

Used properly, you should find your object or a reasonable comparable object on the list. There is something about human nature that wants to believe that what they have is special and/or different from what everyone else has. Do yourself a favor. Think similarity, not differences when looking for comparables within Hopalong Cassidy licensed merchandise. Chances are your item is not as scarce, unusual, or desirable as you think. In fact, it is highly likely that is quite common.

I take full responsibility for the prices that appear below. They are based on the following criteria: (1) what I pay for objects, (2) what I see others pay for objects, (2) prices sellers have marked on objects, (4) prices realized at auction, and (5) what I believe in my heart to be right. I realize that the last point is totally subject, but, at least, I admit it. Blame no one other than me if you do not agree with them.

Hoppy licensed objects will continue to sell at prices far above and below those that appear in this book. This is the why this is only a price **guide**. One of the tasks of Rinker Enterprises, Inc., my business, is tracking and reporting what happens in the antiques and collectibles market. Obviously, I have more than a professional interest in Hoppy items. I welcome learning about any Hoppy objects that you bought, in what condition you bought them, and what you paid for them. This will not be the last Hoppy price list that I prepared. If you are inclined to share your purchase information with me, write to: Harry L. Rinker; Rinker Enterprises, Inc.; 4093 Vera Cruz Road; Emmaus, PA 18049.

Note: Hopalong Cassidy reproductions, copycats, fantasy items, and fakes do not appear on this list. The list includes only items licensed by William Boyd during his lifetime. It is my opinion that reproductions, copycats, and fakes should have minimal value, especially upon resale. The resale value for fantasy items in the secondary market is highly speculative. Time has proven that twenty to thirty years out, most sell for dimes on the dollar. Buy them because you like them, not because of any investment potential.

If you do not find your Hopalong Cassidy item on this list and it looks relatively new, think reproduction, copycat, fantasy, or fake. You will be able to confirm some of your suspicions by checking the reproductions, copycats, fantasies, and fakes illustrated in this book. Keep in mind that new examples are constantly entering the market. To be aware, you must beware. Do not forget.

A Reminder: The following prices are for objects in C6 (very good condition) and without their original packaging.

Advertising Broadside	
Colorado Sage Aftershave, Hoppy Would Have Loved It!	75.00
Dr. West's Dental Kit, 8" x 10", faux wood grain with red and black accents, black and white Hoppy photograph	85.00
U.S. Time, Hoppy "Birthday Gift" watch, 6" x 24"	150.00
Alarm Clock, US Time, 5" diameter, 5 1/2" tall	250.00
Arcade (Exhibit) Card	
Single image of Hoppy	10.00
Hoppy with Topper	10.00
Quintet Card, four images, Hoppy one of four	8.00
Autograph	
Beware of the large number of photographs and other paper documents **with Hoppy's signature printed on them**	
Clipped signature	75.00
Document signed	200.00
Program or similar item signed	125.00
Signed Photograph	175.00
Typed letter signed	250.00
Autograph Book, vinyl leatherette, 5" x 6", zippered on three sides	85.00
Badge	
1 1/4" celluloid button of Hoppy, green ribbon, metal horseshoe /horse hanger	30.00
1 1/4" celluloid button of Hoppy, green ribbon, metal pistol hanger	25.00
Sheriff, silvered brass star, raised portrait image	
Badge	35.00
Badge on display card	75.00
Bank	
Plastic, bronze, 4 1/2" high, issued as part of Hoppy Savings program	30.00
Plastic, Hoppy Statuette Bank dark blue, sold in stores	
Bank alone	50.00
Bank with original box	85.00
Barrette	
Individual unit	15.00
Mounted on display card	45.00
Bath Mat, 48"x 24", chenille, white or light brown ground, Topper image and "Hopalong Cassidy," red and brown lettering	60.00
Bath Towel, 18" x 32 1/2", white terry cloth	75.00

Bedroom Suite, maple, ten piece unit	
Complete Unit	5,000.00
Bunk Bed Unit, two bunk beds, eight wood ends for posts on head and foot boards, four aluminum cylinders for joining beds, wood sign "To Hoppy Ranch" for on top of beds	1,500.00
Chest of Drawers, 32 3/4" x 35 1/2" x 18 3/4"	750.00
Desk, slant front, 40" x 30 3/4" x 18"	750.00
Desk Chair, no Hoppy identification	50.00
Mirror, 27" x 27 1/2", no Hoppy identification	100.00
Ladder for bunk beds, no Hoppy identification	50.00
Night Stand, 23" x 24" x 15", no Hoppy identification	250.00
Toy Chest, no Hoppy identification	250.00
Bedspread	
Chenille, 106" x 90", Hoppy and Topper jumping fence, "Hopalong Cassidy" on bottom	
White ground	100.00
Yellow ground	150.00
Cotton	275.00
Belt, leather, 13"	
Belt alone, Western images on belt readable	100.00
Belt on card	250.00
Bib Overalls, toddlers, cotton twill, gray, all over navy Hoppy and gun images	75.00
Binoculars, decals must be in very good or better condition	
Black	35.00
Gray and red	45.00
Box	75.00
Bicycle, "Hopalong Cassidy Cowboy Bike," Rollfast	
Boy's Version	
Bike only	1,750.00
Full appointments	2,500.00
Girl's Version	
Bike Only	1,500.00
Full appointments	2,000.00
Birthday Card	
4 1/2" x 5 1/2", textured paper, full color illustration, inscription "Well, By Gum! It's Hoppy!," diecut	

opening on right panel with unopened Hoppy Gum Picture
Card, Buzza Cardozo, early 1950s 30.00
5 1/4" x 6 1/4", mechanical, diecut, full color illustration, disk wheel 20.00
6 1/2" x 7 1/2" diecut card with slot, 3 x 7" diecut pistol, inscriptions "Shoot
The Works! It's Your Birthday!," and "Put Your Belt Through The Slits and
"Wear The Gun And Holster Just Like Hoppy," Buzza Cardozo,
Hollywood, early 1950s 35.00
"Here's A Double Barreled Birthday Wish," holster shape,
glass jewel on each gun 25.00
"You're The Big Wheel Today, It's Your Birthday!," mechanical,
four scenes shown by turning wheel, Buzza Cardozo 20.00
Blotter, 4" x 9 1/2", cardboard, American Film Co., Philadelphia,
lists six Hoppy films, red, white, and blue print, blue photos 30.00
Book, Children's, American
A Television Book of Hopalong Cassidy and His Young Friend Danny,
Bonnie Book 35.00
Hopalong Cassidy and Lucky at the Double X Ranch, Garden City
Publishing, pop-up 45.00
Hopalong Cassidy and the Bar 20 Cowboy, Little Golden Book, 1952 20.00
Hopalong Cassidy and the Stolen Treasure, Samuel Lowe Co. 20.00
Hopalong Cassidy and the Two Young Cowboys, Doubleday & Co. 25.00
Hopalong Cassidy Lends a Helping Hand, Bonnie Book, Samuel Lowe, 1950 20.00
Book, Children's, English
Hopalong Cassidy and the Mesquite Gang, Purnell & Sons, Ltd., pop-out
adventure scenes, early 1950s 75.00
Hopalong Cassidy Stories, four annuals, each 40.00
Book, Mulford novel, hard cover,
A.C. McClurg & Company
First edition, first printing, dust jacket 45.00
First edition, first printing, no dust jacket 35.00
First edition, later printings, no dust jacket 20.00
Doubleday
First edition, first printing, dust jacket 40.00
First edition, first printing, no dust jacket 20.00
First edition, later printings, no dust jacket 15.00
A. L. Burt imprint 15.00
Grosset & Dunlap imprint 15.00
Grosset & Dunlap, 1950 retitled 20.00
Garden City imprint 15.00
English edition 25.00
Foreign language edition 35.00
Sample Prices (Auctions, sales lists, etc.)
Hopalong Cassidy Bar 20 Rides Again, Garden City Publishing Co., 1950 15.00
Hopalong Cassidy Sees Red, Grosset & Dunlap, 1921, color dust jacket 35.00
The Coming of Hopalong Cassidy, fifth edition, Grosset & Dunlop 20.00
Book, Tex Burns (Louis L'Amour)
Beware: These novels were reprinted in the 1980s as Tex Burns novels and
in the 1990s as Louis L'Amour novels
Hard cover, first edition 40.00
Soft cover, first edition 30.00
English, Hodder &Soughton 40.00
Foreign language, hard cover or soft cover 50.00
Book Cover, brown paper, back with Hoppy's 10-point Creed
For School Children 25.00
Booklet
Hopalong Cassidy And The Stolen Treasure, Samuel Lowe Co.,
copyright 1950, 48 pages 15.00
Hopalong Cassidy/Trouble Shooter, Samuel Lowe Co.,
copyright 1950, 48 pages 15.00
Boots, black and white leather, white trim, two boot loops,
one with "Hopalong Cassidy," other with "Acme Cowboy Boots"
Pair of Boots 150.00
Boxed 300.00
Boots, rain, rubber, black, white edging and spurs, Hoppy bust on each side, pair
No spurs 50.00
Spurs 75.00
Bottle, Grape Juice, 9 1/4" h, clear, embossed Hoppy portraits and
grape clusters, marked"Cella Vineyards of Reedley, Cal." 35.00
Bow Tie, clip-on, western style, Hoppy and Topper illustration 30.00
Bowl, ceramic, 5 1/4" d, white, Hoppy and Topper color image,
two different colors 50.00
Box
Butter, one pound, generic 35.00
Clothing, generic 250.00
Hopalong Cassidy's Sugar Cones, 5" x 8" x 2", Hoppy art
on front, back, sides, and top 450.00
Ice Cream
Pint, generic 35.00
Quart, generic 55.00
Bracelet
Plastic, black, printed Hoppy image, snap closure 35.00
Sterling Silver, Identification, 9" long
Loose 65.00
Boxed 125.00
Bread End Label Album, Bond Bread, Hopalong Cassidy
Hang-Up Album Hoppy Saves the Gold Shipment
/ Hopalong Cassidy's Creed for School Children 75.00

Hoppy Captures the Stagecoach Bandits / Hopalong
Cassidy's Safety Creed 100.00
Hoppy Captures the Bank Robber's / Hopalong Cassidy's
Creed for Good Manners 125.00
Bread End Label, Bond Bread
Generic Scene 2.00
Label for Hang-Up Album 4.00
Bread Loaf Wrapper
Generic 35.00
Langendorf (recently discovered hoard, several varieties) 25.00
Bubble Gum Card, photographic images from North Vine films
Dangerous Venture, dark blue tone, cards 1 through 23
Borrowed Trouble, brown tone, cards 24 through 47
Hoppy Holiday, red tone, cards 48 through 71
False Paradise, green tone, cards 72 through 95
Unexpected Guests, black tone, cards 96 through 117
Devil's Playground, light blue tone, cards 118 through 141
Fool's Gold, cards 142 through 165
The Dead Don't Dream, cards 166 through 186
Silent Conflict, multi-colored, cards 187 through 208
Sinister Journey, multi-colored, cards 209 through 230
Full set, 238 cards (includes 8 foil cards) 500.00
Individual card
Cards, 1 through 186, single color cards, per card 1.00
Cards, 187 through 230, multi-colored cards, per card 2.00
Foil Cards 15.00
Bubble Gum Card Box 200.00
Bubble Gum Display Sign for 5 Cent Box 50.00
Bubble Gum Display Sign, Store 125.00
Bubble Gum Card Wrappers
1 cent, white 35.00
1 cent, green 50.00
5 cent, yellow 100.00
5 cent, green 125.00
Bunkhouse Clothes Corral, wood, 23 1/2" wide, dairy industry promotion 100.00
Calendar, image of Hoppy on top, advertising in middle, calendar below
1952, full set of 12 sheets 150.00
1952, only December 125.00
Camera
Model HC 310
Camera 50.00
Camera, boxed 85.00
Flash attachment with silk screened image 60.00
Flash attachment boxed 100.00
Candy, Hoppy Pops, Topps
Box 50.00
Wrapper 20.00
Candy Bar Box, 8" x 8 3/4" x 5", Hopalong Cassidy Chocolate
Coconut Candy Ryan Candy Co., Ltd., NYC, red, black
lettering and Hoppy image 150.00
Candy Bar Wrapper, Hopalong Cassidy Chocolate Coconut Candy,
5 x 6", foil, black and white design, orange accent 65.00
Cap
Fabric, brim, color portrait of Hoppy 50.00
Wool, winter pull over, Hoppy on Topper and text on turned
up section, Pedigree Sportswear 75.00
Cartoon Art, original
Comic Book Page 200.00
Daily Comic Feature 150.00
Sunday Comic Feature 300.00
Catalog, *Bicycle Journal,* October, 1950, black and white
Hoppy tricycle advertisement on front cover, 48 pages, 6" x 9" 50.00
Cereal Bowl
Ceramic, two different decals 45.00
Glass 40.00
Cereal Box
Post's Raisin Bran, Hoppy badge premium 75.00
Wheaties, back panel including "Trail Dust" movie image 125.00
Chair, child's, television, 22" x 16 1/2" x 12 1/2", wood frame, folding,
black material back and seat, white fringe dec, Hoppy on bronco
illustration on seat, marked "Hopalong Cassidy Official Bar 20
TV Chair" on back rest 400.00
Chap and Vest Set, suede leather chaps and vest, possibly from ensemble set 150.00
Chow Set, Bar-20 Chow Set, white milk glass, bowl, plate, and tumbler
Bowl 40.00
Plate 40.00
Tumbler 40.00
Boxed set 200.00
Clothes Hamper, 23" x 16" x 9 1/2", metal, red, Hoppy and
Topper illustration 175.00
Coaster, 4" d, "Spun Honey Spreads like Butter"
advertisement, fake No Real Value
Coin, 1 1/8" d, metal, Hoppy head and "Hopalong Cassidy-William
Boyd" reverse with horseshoe and clover and "Good Luck from
Hoppy", 1 1/8" diameter, has been reproduced, price is for period piece 10.00

Cole Brothers Circus
  Broadside, Cole Bros. Circus Presents "In Person" William
    Boyd / HOPALONG CASSIDY, 17" x 26", late 1940s   250.00
  Program, Hoppy's picture on the cover   75.00
Coloring Book
  Abbott Publishing, William Boyd cover   40.00
  Doubleday   45.00
  Doubleday, corporate sponsored   50.00
  Samuel Lowe Co., William Boyd cover   20.00
  Whitman   35.00
Coloring Outfit, Transogram   150.00
Comic Book
  William Boyd Western
    Number 1   65.00
    Number 2   40.00
    Numbers 3 through 10   25.00
    Numbers 11 through 23   20.00
  Hopalong Cassidy
    Number 1   500.00
    Number 2   125.00
    Numbers 3 to 5   75.00
    Number 6 to 10   35.00
    Number 11 through 25   25.00
    Number 26 through 50   17.50
    Number 51 through 85   10.00
    Number 86, Fawcett first issue   50.00
    Number 87   35.00
    Numbers 88 through 108   15.00
    Numbers 109 through 135   10.00
    Foreign language   20.00
  Master Comics   25.00
  Promotion
    Bond Bread   25.00
    Grape Nut Flakes   20.00
    White Tower   25.00
  Real Western Hero   50.00
  Six Gun Hero   25.00
Cookie Box, Hopalong Cassidy Cookies, Burry, 5" x 8" x 2"   250.00
Cookie Jar, lid must be present 5" high, 8" diameter, ceramic,
  "Cookie Corral," brown ground, raised design, Hoppy decal on front   250.00
Cookie Jar, 11" h, 7" d, ceramic, "Cookie Barrel," white ground,
  Hoppy decal on front, saddle shape finial, green cactus motif around bottom
  350.00
Cowboy Ensemble (Outfit)
  Boys vest and chaps, black, white trim, Herman Isken Co., boxed   150.00
  Girls vest and fringed skirt, black, white fringe, color image,
    Herman Isken Co., boxed   150.00
  Girls vest and fringed skirt, black with white piping, attached
    silver studs and red glass stone metal cowboy boots and
    hat pins on vest, buckles on sides of waist, boxed   200.00
  Individual pieces of clothing from boxed sets   50.00
Cottage Cheese Container, generic, 12 oz., 4 1/2" diameter   25.00
Crayon and Stencil Set, Transogram Toys   125.00
Crayons [Note: There was no crayon set. These are loose pieces
  from the Crayon and Stencil set]   No Real Value
Decals, watercolor Hoppy and Topper transfers, set of three, unused, c1950   25.00
Dental Kit
  Complete Unit   150.00
  Mirror   20.00
  Toothbrush   25.00
  Toothbrush in its box   40.00
  Tube of Toothpaste   25.00
  Tube of Toothpaste in its box   40.00
Dinner Set, 9 1/2" d plate, 5 1/4" d bowl, 3" h mug, china, full color
  illustration, original cardboard box, W. S. George Pottery Co.,
  East Palestine, Ohio   200.00
Dixie Lid
  2 1/4" d, rim inscription "Staring in Harry Sherman Production"   20.00
  2 1/4" d, rim inscription "Featured in Harry Sherman Production"   20.00
  2 3/4" d, rim inscription with Hoppy name and film title, late 1940s   15.00
Doll
  20 1/2", stuffed cloth, soft vinyl head and hands, removable black
    outfit, attached black, white, and green cello button   300.00
  22" h, Ideal, stuffed, plastic head, plush hair and chaps, two plastic
    holsters, two metal guns
    Doll   200.00
    Doll with cardboard box with paper end label   250.00
  30" h, cloth body, plastic face, black outfit, removable black felt hat   300.00
Dominoes, Milton Bradley, cardboard, three unpunched sheets,
  each with Hoppy images   50.00
Fabric Remnant
  Bust portrait of Hoppy, signature, assorted western scenes, green ground 100.00
  Bust portrait of Hoppy, black on white ground   75.00
Figure
  Advertising, U.S. Time, plaster, three dimensional figure of Hoppy   750.00
  Carnival, chalkware, 13 1/2" high, standing Hoppy   250.00
  Cast Metal, Hoppy with drawn guns, black, green base, 3" high,

  possibly fantasy piece   10.00
  Ideal Toy Company, plastic, three dimensional Hoppy and Topper
    Figure, complete   100.00
    Figure, complete, with box   150.00
  Set, lead, The Official Hopalong Cassidy Western Series, English,
    Timpo, seven figures, each 2 1/4" high
    Individual figure   35.00
    Set   350.00
Figure and Paint Set, Hopalong Cassidy and His Pals Lucky
  and California, Laurel Ann Arts, Los Angeles, ca. 1950
    Unpainted   400.00
    Painted figures   250.00
  Film Strip set, "Heart of the West," six filmstrips and
    plastic viewer in box   150.00
Flashlight, red band with color Hoppy image, Morse code listing on back   60.00
Flashlight Gun, 8 1/2" l, black hard plastic, white Hoppy name on
  sides, Bar 20 Ranch symbol grip sides, raised black steer head,
  barrel end with flashlight bulb
    Loose   75.00
    Boxed   150.00
Game
  Chinese Checkers, Milton Bradley cardboard, sheriff's
    badge playing field, marbles   75.00
  Hopalong Cassidy Game, Milton Bradley, playing
    board, money, cardboard coins, metal figures, and cardboard spinner   50.00
Game, Card, Hopalong Cassidy Canasta, Pacific Playing Card Company
  Full set - plastic saddle tray, cards, and box   150.00
  Plastic Saddle   35.00
  Playing Cards   25.00
Game, Toss
  Lasso Game, Transogram, 12 1/2" x 15 1/2", plastic target
    figure of Hoppy on Topper, four rings   150.00
  Pony Express Toss Game, Transogram, 12" x 18",
    bean bags masonite target   65.00
Garden Set, Hopalong Cassidy Bar 20 Chuckwagon Garden, Topper's
  feedbag, four growing mats, paper cut-out ranch buildings and fence,
  directions, signed image of Hoppy, original box   250.00
Glass
  3", bathroom, white milk glass, "Healthy Pals!" and "Mornin' Pards!"   50.00
  3", juice, milk glass, Hoppy and Topper illustration, black writing,
    "Healthy Pals!," on reverse "Lasso Onto Sunshine, Rope Yourself
    a smile, Be like Hop and Topper, Make each day worthwhile!"   35.00
  3", mug, white milk glass, set of four, each with different illustration
    and color
    Set   75.00
    Single   15.00
  4 3/4", Western Series, clear glass, decal motif, at least four known, each   35.00
  5", milk glass, tumbler style
    Breakfast   25.00
    Lunch   25.00
    Dinner   25.00
Gloves, leather, fringed, wrist guard, with steer head conch,
  white stamped "Hopalong Cassidy"   75.00
Greeting Card
  Christmas, "Hi There!" and "Hoppy Bar 20 Ranch," Topper,
    Hoppy and Topper at fence, Buzza Cardozo   17.50
  Christmas, "Here's Santa and Hoppy With a Greeting For You,"
    Hoppy and Santa, Buzza Cardozo   17.50
  Get Well, 5 1/4" x 6 1/4", full color illustration of Hoppy and
    Topper, printed game inside, marked "Official Hopalong
    Cassidy Cards" and Buzza Cardozo of Hollywood on back   25.00
Guitar, black and tan, Hoppy illustration, Jefferson Mfg. Co.,
  fantasy item   No Real Value
Gun
  Cap Pistol
    Authentic Shoot 'n Iron, Australian, boxed   175.00
    7", brass finished, black plastic grips each with name spelled in white rope
      script   75.00
    7 1/2" long, silvered metal, ivory plastic grips each with name in rope script
      75.00
    9", gold finished, black plastic grips each with etched portrait
      Gun alone   125.00
      Gun with black leather holster   200.00
    9" long, silvered metal, ivory plastic grips each with etched portrait   100.00
  Paper Pistol, stiff cardboard, click, Dairylea premium   75.00
  Paper Pistol, stiff cardboard, pop
    7" long, black and white, red Hoppy name on barrel,
      Buzza Cardozo, early 1950s   35.00
    9 1/2", Capitol Records premium, early 1950s   65.00
  Rifle, Hop A' Long, Marx, tin, clicker, 30", colorful decal on butt   250.00
  Set   *see Holster*
Gym Set, Hopalong Cassidy 5-Piece Gym, rubber muscle builder,
  two dumbbells and guns, original box and directions   225.00
Hair Bow, 3 1/4" x 4 3/4", yellow cloth, printed Hoppy images, period card   65.00
Hair Trainer
  4 oz. bottle   30.00
  Quart bottle   75.00

Handkerchief, 11 1/2" x 11 1/2", cotton, white, stitched inscription
  and portrait, blue "Hopalong Cassidy" and red "Topper" name
  Handkerchief only — 65.00
  Handkerchief with accompanying tin steer head slide — 100.00
Handbill
  Grocery, Grape Nut Flakes, 8 1/2" x 11", promotes Hoppy TV
    show, probably grocery bag insert — 75.00
  Movie, 8 1/2 x 11", paper, blue, black print,
    Hoppy and masked bandit illustration — 45.00
Hat, black felt, white neck string, leather slider printed on front
  "Hopalong Cassidy" and "Deputy," steer head illustration — 150.00
Holster
  Black leather holster with red glass beads, two black grip guns with
    Hoppy engravings, gold barrels, "Hopalong" on reverse of belt — 125.00
  Black leather holster with silver accents, holds six silver wood bullets
    Loose — 100.00
    Boxed — 150.00
  Set, black leather belt, double holsters, pair of gold finished metal
    cap guns, black plastic grips
    Loose — 400.00
    Boxed — 600.00
Hunting Knife, 4" l, miniature, plastic and steel, 5" black vinyl belt
  loop scabbard marked "Hopalong Cassidy," knife handle marked
  "USA", modern fantasy item — No Real Value
Ice Cream popsicle Bag, 5 1/4" x 13", Dairylea Ice Cream,
  insulated foil, Hoppy photo and ice cream name — 50.00
Ice Cream Container
  Pint, generic — 50.00
  Quart, generic — 65.00
Indian Head Band, "Big Chief Hoppy," black feathers, Hoppy
  bust illustration on band — 75.00
Inflatable Toy, vinyl, Topper, 18" x 19" — 100.00
Invitation, 3 1/4" x 4", full color, black and white Hoppy photo,
  Buzza Cardozo, Hollywood, original envelope — 15.00
Jacket, denim, black, metal snaps with "Hopalong Cassidy," white
  trim, front pockets with steer head monograms, original label, Blue Bell — 75.00
Jelly Jar, clear glass, Hoppy and Topper illustration — 25.00
Label, 4" x 4 3/4", TV chair, Transogram Co., NYC, Hoppy
  and Topper illustration — 30.00
Lamp
  Holster Night Light — 200.00
  Night Light, 7 1/2" tall, white glass cylinder, color decal on front,
    same base as table lamp — 250.00
  Table, cylindrical base, decal
    With shade — 400.00
    Without shade — 200.00
  Table, horse head top base, bottom portion of base red, nothing on base
    indicates that the lamp was a Hoppy product. It is possible that some
    dealer added a Hoppy shade and started the legend
    With shade — 400.00
    Without shade — 250.00
  Econolite Hopalong Cassidy Bar-20 Ranch — 275.00
Leaflet, 3 1/2" x 5 1/4", blue and white advertisement for Heart
  Of The West Fox Theater, LaRue, Ohio, 1945 — 25.00
Lobby Card
  Sherman Hoppy
    Title Card — 25.00
    Standard Card — 15.00
    Eight card set — 150.00
  William Boyd Hoppy
    Title Card — 20.00
    Standard Card — 10.00
    Eight Card Set — 125.00
Lunch Kit
  Metal, red or blue, irregular decal
    Box — 85.00
    Thermos — 45.00
  Metal, red or blue, rectangular decal
    Box — 100.00
    Thermos — 50.00
  Metal, lithographed scene
    Box — 150.00
    Thermos — 75.00
Magazine
  Hoppy on cover, generic — 30.00
  Article about Hoppy inside — 15.00
  Sample Prices (Auctions, sales lists, etc.)
    Friends, August, 1950, 20 pages, distributed by Chevrolet
      dealers, 10 x 13 1/2" — 30.00
    International Photographer, January, 1937, Hoppy on Topper
      cover, 8 1/2 x 12" — 40.00
    Life, June 12, 1950, Hoppy cover, three page article with photos — 20.00
    Look, August, 1950, Hoppy and boy on cover, feature story
      about Cowboy craze — 30.00
    Quick, May 1, 1950, bust of photo of Hoppy on cover, two page article — 30.00
    Time, November 27, 1950, Hoppy cover and article — 25.00
    TV Digest, November 29, 1952, Hoppy cover — 85.00

Tele-Views, February 1, 1951, Hoppy cover — 75.00
Magazine Tear Sheet
  Acme Western Boots, Saturday Evening Post, full page color ad
    Matted — 25.00
    Unmatted — 10.00
  Burnett's Instant Pudding, Life, full page, unmatted — 12.00
  Grape-Nuts Flakes, full page, unmatted — 12.00
Mask
  Latex, face
    Mask alone — 75.00
    Boxed — 150.00
  Paper, face, 13" wide, 11" high, advertising on back for pudding — 65.00
Marble, black, Hoppy image, fantasy item — No Real Value
Marionette, Hopalong Cassidy Stringless Marionette, National
  Mask & Puppet Corp., plastic head, hands, and feet, black fabric body — 275.00
Milk Bottle
  Pint, generic — 90.00
  Quart, generic — 75.00
  Half gallon, generic — 125.00
  Sample Prices (Auctions, price lists, etc.)
    Cloverdale, quart — 80.00
    Dairylea, pint — 70.00
    Northland Dairy, Des Moines, Iowa, 9" h, quart, red
      Hoppy illustration — 100.00
    Producers Dairy Delivery, Inc., Fresno, CA, 10 3/4" h, 1/2 gal,
      red printing, Hoppy image and "Hoppy's Favorite Milk"
      [Note: This dairy still holds a Hoppy license.] — 75.00
Milk Bottle Cap, generic — 15.00
Milk Container, waxed cardboard
  Generic, period — 85.00
  Producers, all current production, each — 5.00
    1/2 pint, Hoppy head, red and white
    Pint, Hoppy on Topper, red and white
    Quart, full length Hoppy, red and white
    Quart, full length Hoppy, pink and white
    1/2 gal, Hoppy on Topper, red and white
    1/2 gal, full length Hoppy, red and white
    1/2 gal, chocolate milk, Hoppy on Topper, yellow and brown
Movie Film
  8mm, black and white, silent — 25.00
  16mm, black and white — 30.00
  Titles in series
    Bar 20 Rides Again
    Ghost Canyon Roundup
    Heart Of The West
    Law of the Trigger
    Stick To Your Guns
Movie Poster
  William Boyd, pre-Hoppy
    One-sheet, Boyd major star — 250.00
    One-sheet, Boyd minor star — 125.00
  Sherman Hoppy
    One-sheet, original release — 250.00
    One sheet, re-release — 150.00
    Three-sheet — 450.00
  William Boyd Production Hoppy
    One-sheet, original release — 100.00
    One-sheet, re-release — 65.00
    One -sheet, retitled re-release — 50.00
  Foreign Release
    Sherman Hoppy, one-sheet — 175.00
    William Boyd Hoppy, one-sheet — 50.00
    Re-release, one sheet — 40.00
  Sample Prices (Auction, sales lists, etc.)
    Bar 20 Justice, Paramount, 1938, Hoppy, Gabby Hayes, and Russell Hayden
      150.00
    Burning Gold, Republic, 1936 — 80.00
    Dangerous Venture, United Artists, 1947, Boyd and Andy Clyde — 50.00
    Dress Parade, Pathe, 1927, Boyd and Bessie Love — 250.00
    False Colors, full color pastel, marked "released through Variety Film
      Distributors" — 75.00
    Flaming Gold, RKO, 1933, Boyd and pat O'Brien — 175.00
    Fool's Gold, Untied Artists, 1947, Boyd and Andy Clyde — 45.00
    Heart of Arizona, 41" x 81", full color, Hoppy and Gabby Hayes — 150.00
    Hills Of Old Wyoming, Paramount, 1937, Boyd and Gabby Hayes — 80.00
    Hopalong Cassidy Returns, 41" x 81", full color Morgan lithograph — 100.00
    Man From Butte, 27" x 37", Mexican re-issue — 55.00
    Mystery Man, color, overprinted "Country of origin, USA" — 250.00
    Riders of the Deadline, 28" x 44", full color, Variety Film Distributors — 75.00
    Rustler's Valley, 1937 — 175.00
    Song Of The Volga Boatman, 1926 — 200.00
    The Dead Don't Dream, 25" x 31", half sheet, full color,
      black wood frame — 100.00
    The Fighting Cowboy, Spanish print — 180.00
    The Marauders, 41" x 81", re-issue, 1947 — 65.00
    Trail Dust, Paramount, 1936 — 175.00

Movie Viewer, Acme Plastics, Inc., Miami, FL, #15, 3" x 3/4" x 1 3/8"
  viewer and two Hoppy films, films in green boxes, c1955 75.00
Mug, 3" high, white glass, red picture and signature 15.00
Neckerchief, 36" long, silver plastic tie slide 65.00
Neckerchief slide, metal, steer's head, red glass eyes 25.00
Necktie, child's, 13" long, blue, gray, white, red, and yellow portrait, white
  inscription, pre-tied, adjustable 50.00
Notebook, two ring binder 75.00
Notebook Filler Paper, 24 three hole sheets, original packaging sleeve
  with color Hoppy and Topper illustration 35.00
Paint Set, *see coloring outfit*
Paperback Mulford novel
  American Services Edition 25.00
  Dell 20.00
  Pocket Books 15.00
  Popular 20.00
  Western Novel Classics 25.00
Paper Cup, 5 1/2" h, full color images of Hoppy on Topper, set of six,
  mint in package 60.00
Paper Money, 2 3/4" x 6 7/8", Joe Fredrickson Toy & Game Co., "Lucky
  Bucks" and Hoppy image, three denominations -5, 10, and 100, each 2.00
Paper Napkin, color Hoppy and Topper image, set of 8, mint in package 50.00
Paper Plates, full color images of Hoppy on Topper, set of six, mint
  in package 60.00
Paper Tablecloth, 53 1/2" x 88", mint in package 75.00
Party Kit, Hopalong Cassidy Official Party Kit, six invitations and
  reply cards, three sign decorations, twelve placemats, Pin The Tail
  On Topper game, and Official Party Kit Booklet, original box,
  Buzza Cardozo, 1950s 200.00
Pen, 5 3/4" l, black and silver, "Hopalong Cassidy" printed on side,
  top with figural Hoppy image, ball-point
  Pen alone 50.00
  Pen boxed 75.00
  Refill with box 25.00
Pencil Case
  3" x 8", stiff paper, faux wood grain finish, black and white photo,
    PA shoe store advertisement stamped on back 60.00
  3 3/4" x 9", vinyl, red, Hoppy image in center, zipper top, blue
    pencil, eraser, and plastic pen shaft 75.00
  4 1/2" x 8", plastic, gun shape, blue, black and white
    Hoppy on Topper illustration, three pencils, five crayons, eraser,
    and pen shaft 100.00
  4 1/2" x 8", plastic, gun shape, red, plastic, black and white Hoppy
    on Topper illustration, holster slip case, three pencils, five crayons,
    eraser, and pen shaft 125.00
  5" x 8 1/2", cardboard, tan, black and white portrait and inscriptions,
    compartment int. with sliding drawer, metal snap fastener 90.00
Pencil Sharpener, 2 5/16" x 1 15/16" x 1/2", plastic, red, white, and
  black, Hasbro 20.00
Pennant
  20" l, black felt, white trim, illustration and script 25.00
  27 1/2", black felt, white trim, illustration and script 35.00
Photo
  3 1/4" x 5 1/2", Bond Bread fan photo, full color, glossy, stamped on
    reverse "See Hoppy on Television Every Saturday 4 to 5 P.M. WNBK
    Channel 4" 15.00
  5" x 7", tinted, black and white thin photo paper, waist portrait of
    Hoppy, from dime store picture frame, early 1940s 20.00
  8" x 10", black and white, movie stills, glossy
    Bar 20 Rides Again scene 5.00
    Hoppy and Robert Mitchum 7.00
    Hoppy cleaning gun 8.00
    Hoppy, Topper, and heroine 8.00
    Hoppy with gun drawn 10.00
  8" x 10", tinted, Hoppy sitting on Topper, dime store frame, early 1950s 40.00
Photo Album, 10" x 14", tan vinyl, embossed full color portrait of Hoppy
  on Topper on cov, black pages, early 1950s 100.00
Picture Frame, personalized, wood, 6" x 10", department store photograph
  of Hoppy and youngster, early 1950s 75.00
Picture Frame, Stock
  Dime store, cardboard, black and white with gold accents, 7" x 9 1/4",
    black and white promo picture 50.00
  Dime store, metal, publicity photograph of Hoppy sitting on Topper 40.00
Picture Gun and Theater, 6" gray metal gun, two boxes each containing
  six films, cardboard fold out screen, box
  Compete set 125.00
  Gun 40.00
  Box of film 25.00
Pillow Case, satin
  21" x 20", gold trim and fringe, yellow, red, blue, and black illustration
    of Hoppy and Topper, inscribed "Hopalong Cassidy" 75.00
  21" x 21", gold trim and fringe, black, rose, blue, and green illustration
    of Hoppy 75.00
Pinback Button
  1 3/8" diameter, orange, Hoppy and Topper, "Hopalong Cassidy's
    Saving Rodeo" 20.00

1 3/8" diameter, red and yellow, full length Hoppy, "Hopalong
  Cassidy Daily in the Chicago Tribune" 25.00
Pistol and Spur Set, two spurs with straps, one pistol, box 375.00
Placemat, 12" x 18", vinyl, full color illustration, red border, tan rope script 50.00
Plate
  9 1/4" d, ceramic, Hoppy and Topper in blue, yellow, and green 45.00
  9 3/8" d, ceramic, Hoppy and Topper profile in black and red 45.00
Pocket Knife
  2 1/4" 35.00
  3 1/2" 50.00
  Plastic belt loop, add 10.00
Pocket Protector, 2 3/4" x 5 1/2", plastic, holds salesman's profit chart,
  dark brown, gold Hoppy on Topper illustration, Movieland Girl From
  Hollywood knit underwear on reverse, c1950 50.00
Pogo Stick, 46 1/2" tall, steel rod, spring action 200.00
Post Card
  American
    Endorsement Card, 3 1/4" x 5 1/4", car dealer's, black and white
      glossy photo of Hoppy posed at door of station wagon, reverse
      inscription "Hopalong Chooses The New 1942 Chrysler Town And
      Country Station Wagon As His Personal Car" 17.50
    "Home of Hopalong Cassidy," color, Boyd's picture in upper right
      corner, 3 1/2" x 5", early 1950s 35.00
    Promotional, Hoppy holding large cartoon magazine, 3 1/2" x 5 1/2" 15.00
  English
    Hopalong Cassidy Series, color, 3 1/2" x 5" 20.00
Potato Chip Can, 7 1/2" diameter, 11 1/2" high 150.00
Premium
  Folder, generic, identifies Hoppy premiums used for dairy
    industry promotions 75.00
  Photo, 7" x 8 1/2", black and white glossy, Barclay Knitwear Co.,
    1949 Boyd copyright 25.00
  "Ranch House Race," game, thin cardboard, 10 3/4" x 13 3/4",
    sponsored by Stroehman's Sunbeam Bread 100.00
Pressbook
  Sherman Hoppy 45.00
  William Boyd Hoppy 25.00
  Hopalong Cassidy Publicity and Exploitation Manual, early 1940s 50.00
Print Publicity card, Paramount Pictures, 5" x 7", waist portrait of
  Hoppy holding pair of guns, early 1940s 20.00
Promotion Book, *Hopalong Cassidy Corrals Everybody*, lists
  merchandise, 28 pages, 1949 50.00
Pulp, Hopalong Cassidy's Western Magazine, Best Books, Inc.
  Volume 1, Number 1, Fall 1950 150.00
  Volume 1, Number 2, Winter 1951 100.00
Puppet
  10 1/2" h, printed fabric body, molded vinyl head with black hat,
    inscribed name, 1951 copyright 100.00
  Full figure, cloth body, molded vinyl head with black hat 150.00
Puzzle
  Milton Bradley, Hopalong Cassidy Television Puzzles, boxed set of four 60.00
  Milton Bradley, No. 4025, boxed set of three, Hoppy and Topper 50.00
  Whitman, frame tray, 9 1/4" x 11 1/2"
    Frame tray puzzle alone 35.00
    Frame tray puzzle with cardboard sleeve 50.00
Radio, red or black case (assumes wire antenna is present)
  Front panel has Topper's legs up in air 350.00
  Front panel has Topper's legs down 300.00
Raincoat and hat, rubberized, black, Hopalong Cassidy color patch
  on each arm 150.00
Recipe Pamphlet, 6" x 3 1/2", Hoppy's Bar-20 Ranch Recipes,
  Chicken of the Sea advertisement, color photos 35.00
Record Album, storybook, two record set
  Hopalong Cassidy and the Singing Bandit
    45 rpm 45.00
    78 rpm 35.00
  Hopalong Cassidy and the Square Dance Holdup
    45 rpm 45.00
    78 rpm 35.00
  Single
    45 rpm 30.00
    78 rpm 25.00
  Sample prices (Auctions, sales lists, etc.)
    Billy and the Bandit, 78 rpm, Little Folks Favorites label, original sleeve 20.00
    Hopalong Cassidy and the Big Ranch Fire, 45 rpm, Capitol 30.00
    Hopalong Cassidy and the Mail Train Robbery, 45 rpm, Capitol 35.00
    Hoppy's Happy Birthday, 45 rpm, Capitol 35.00
    My Horse Topper, 78 rpm, original sleeve and paper jacket 25.00
    The Story of Topper, 78 rpm, original sleeve and colorful jacket,
      Capitol Records 40.00
Ring
  Plastic, William Boyd decal, decal edges yellowing, 11/16" x 11/16" x 1/2" 15.00
  Silver, adjustable, horseshoe with Hoppy head in center,
    3/4" x 3/4" x 1/2" 20.00
Rocker, child's, silvered tubular steel braces, matching flat steel rockers,
  red seat white back rest, vinyl upholstery, Hoppy and Topper and name
    Rocker with horse head center unit 250.00

Rocker without horse head center unit — 200.00
Rocking Horse, Rich Toys Inc., plastic body, wood legs and rockers, black and white yellow accents on saddle, red rockers and handles — 275.00
Roller Skates, pair, steel, adjustable, cowhide straps, removable metal spurs, skate key and instruction card
  Skates alone — 200.00
  Boxed — 300.00
Rug
  24" x 60", chenille — 75.00
  48" x 77", chenille — 125.00
Saddle, child's, leather — 450.00
School Bag, 10 1/2" x 14", fabric, brown, Hoppy on Topper portrait, white cloth circle with silvered metal star badge inscribed "Hoppy" — 125.00
Scrapbook, 12" x 15", cardboard with tan vinyl, embossed full color portrait, western images, and two brown and white images — 85.00
Sheet Music
  Note: These sheets and albums sell among sheet music collectors in the $5.00 to $8.00 range
  Hoppy song written during Hoppymania craze of early 1950s — 20.00
  Song Folio, Hoppy's picture on cover — 35.00
  Theme Music, Hoppy movie — 15.00
  Sample Prices (Auction, price lists, etc.)
    Bob Atcher's *Meadow Gold Song*, 8" x 10", premium from Chicago Hoppy TV show, back cover has advertisement for Meadow Gold Butter — 100.00
    Lazy Rolls The Rio Grande, blue and white, Hoppy riding Topper, cattle herd background, 1939 — 15.00
    Take Me Back to those Wide Open Spaces, Hoppy on Topper — 25.00
    Where The Cimarron Flows, Harry Sherman, 1939 copyright — 20.00
Shoe Rack
  16" x 14 1/2", plastic, yellow, black and red printed images, eight pockets — 100.00
  19 3/4" x 17 1/2", plastic, red, black trim, yellow and black printed images, eight pockets — 85.00
Shirt, western style, black, long-sleeve, white piping trim, color Hopalong Cassidy monogram, The Little Champ, fabric by Tanbro — 50.00
Shooting Gallery, lithograph tin, 14" x 16" x 5", gun and target toy, wind-up feature, Automatic Toy Company
  Loose — 100.00
  With box — 275.00
Shooting Game, Bar Twenty Shooting Game, English, Chad Valley, die cut cardboard target figures, two metal/wood pistols for firing rubber bands, 1953 — 250.00
Sign
  5 1/2" x 7", die cut, cardboard, hanger, Bond Bread — 200.00
  10 1/4" x 13 1/2", Bond Bread, advertising sponsorship of Hoppy TV Show, window display card — 150.00
  11" x 15 1/2", cardboard, yellow, black printing, open area for name of special guest to appear on Hoppy TV show, sponsored by Wonder Bread — 150.00
Sleeping Bag, 26" wide when rolled, cloth storage bag — 250.00
Soap, four bars, Hoppy image on each bar, original box, trading card removed, missing front cellophane, Dagger & Rambled, Inc., Newark, NJ, box 4 1/2" x 6 1/2" — 85.00
Slate Set, Hopalong Cassidy School Slate Outfit, 9" x 11" wood slate, white side for crayons, black side for chalk, crayons, stencils, chalk, slate pencils, and pictures, original box, Transogram, 1950 — 250.00
Socks, black and white stripe, black ankle band, printed Hoppy image, pair — 40.00
Stationery Set, 10 3/4" x 8 3/8" red, blue, and yellow folder, Hoppy letterhead with blue bust image in top corner, plain white envelopes, Whitman Publishing Co. — 75.00
Straws, boxed set, assorted colors, full color Hoppy illustration — 50.00
Sweater
  Pullover, sleeves — 85.00
  Pullover, sleeveless — 65.00
  Button down — 100.00
Swim Trunks, child's, size 6, cotton, blue, woven Hoppy and Topper illustration on front corner bordered with yellow lasso design — 85.00
Tab, 2 1/4" x 1 1/4", orange and black felt, folding tab with Hoppy head and "Stroehamann's Bread, Hoppy's Favorite" — 30.00
Tablet, School, 8 x 10", full color image on green background, unused — 20.00
Target Game
  Hopalong Cassidy Target Game, Marx, lithograph tin — 175.00
  14" x 17", lithograph tin, full color, Chad Valley Co. Ltd, Harborne, England, c1950 — 200.00
Tie Bar
  Bar, Horseshoe center with "Hopalong Cassidy"
    Tie Bar — 50.00
    Boxed — 100.00
  Bar, clip holds chain, holds broncobuster cowboy on horse, gold colored metal
    Tie Bar — 65.00
    Hoppy name on insert board and box lid — 125.00
Tie Rack, 3 1/2" x 11 1/2", composition wood, raised image of Hoppy and Topper with gold finish, brown raised western motifs, ten metal holders — 85.00
Toiletry Set, Hopalong Cassidy Dudin-Up Kit, shampoo, hair trainer, and plastic comb, original box — 175.00

Toy
  Automatic Television Set, Automatic Toy Co., Staten Island, NY, 5" x 5 3/8" x 5" television, plastic, wind-up, four film strips, original box with Hoppy images
    Loose — 75.00
    With box — 125.00
  Hopalong Cassidy's Western Ranch Playhouse, stiff paper, 1950 copyright, 48" x 34" x 44" assembled — 250.00
  Rocker, Hop "A" Long Cassidy, Marx Range Rider, wind-up, lithograph tin, rocks back and forth, arm and lariat move — 300.00
  Sparker Toy, 3 3/4" x 2" x 1", plastic, bust of Hoppy, black, red and silver accents, protruding, plunger moves flint wheel and sends sparks out sides — 75.00
  Western Frontier Set, Milton Bradley, village street scene, three dimensional
    Complete boxed set — 350.00
    Individual buildings — 40.00
  Zoomerang Gun, plastic, red, period box — 85.00
Toy Chest, wood, Hoppy decal on front, red script "Hopalong Cassidy" on lid — 350.00
Thrift Kit, Hopalong Cassidy Savings Club Thrift Kit, certificate, letter, full color photo sheet, postcard, folder sheet, original envelope — 85.00
Trading Card
  Hopalong Cassidy, Ways of the West, Bond Bread, 16 in set, per card — 3.00
  Hopalong Cassidy Wild West, Post Cereal Premium, 36 in set
    Full set — 125.00
    Card featuring Hoppy's image on front — 4.00
    Card not featuring Hoppy — 2.00
    Advertisement card for set — 3.00
Transfer Strip, 2" x 7 1/2", three full color water transfer of Hoppy posed with Topper, unused, c1950 — 45.00
Troopers Club
  Application Card, 3 1/4" x 5 1/4" — 20.00
  Newsletter — 25.00
Underwear (briefs), boy's, scenes of Hoppy and Topper — 100.00
Utensil Set, Hopalong Cassidy Junior Chow Set, stainless steel, knife, fork, and spoon
  Complete set in box — 125.00
  Individual units — 30.00
Valentine, "I'd Like To Tie Your Heart To Mine," Buzza Cardozo — 30.00
View Master Reel
  #955, original envelope — 10.00
  #956, "The Cattle Rustler," envelope, copyright 1953 — 15.00
Wallet
  3 1/2" x 4 3/8", leather, color Hoppy and Topper image on plastic, metal sawtooth design trim, Bar 20 special agents pass — 55.00
  3 1/2" x 4 1/2", black leather, Hoppy and Topper image on front, reverse with "Good Luck from Hoppy," zipper closure, owner's special agents pass — 75.00
Wallpaper
  Roll — 125.00
  Strip, 17 3/4" w, 48" l, full color Hoppy on Topper and western images — 50.00
Window Curtains
  Chenille, two 35 1/2" x 84" panels, light brown, green, brown, and yellow trim, red and brown lettering, inscribed "Bar 20" and "Hopalong" — 200.00
  Cotton, two 32" x 69" panels, blue, white illustration of Topper's head, inscribed "Bar 20 Ranch" in red, repeated "Hopalong Cassidy", possibly homemade using Hopalong Cassidy printed theme fabric — 125.00
Wood Burning Set, American Toy and Furniture Co., burner, holder, instructions, paint brushes, water color paints, dish, and extra tips — 225.00
Wrist Compass, vinyl strap with silvered plastic case, fig. image of Hoppy — 50.00
Wrist Cuffs, 5 1/2" x 4 3/4" x 1", leather, studded with glass jewels, stamped "Hoppy" in white
  Loose — 75.00
  Boxed — 150.00
Wristwatch, US Time, working order
  Large Size
    No Wrist Band — 45.00
    Wrist Band — 75.00
  Small Size
    No Wrist Band — 40.00
    Wrist Band — 65.00
  Box (add to value of watch with wrist band)
    Flat — 100.00
    Saddle, American — 150.00
    Saddle, English — 200.00